Surface Geology of the Pacific Northwest

Geology Lab Manual

Written and compiled by
Shannon Othus-Gault, Reanna Camp-Witmer,
and Autumn Christensen

with additional contributions from Michelle Harris and Andrew Frank

Surface Geology of the Pacific Northwest: Geology Lab Manual
ISBN: 978-1-955499-41-5
© 2017, 2018, 2025 by Chemeketa Community College.

Chemeketa Press
Chemeketa Community College
4000 Lancaster Dr NE
Salem, Oregon 97305
collegepress@chemeketa.edu
chemeketapress.org

A full list of credits appears on pages 149–150 and constitutes an extension of this copyright page.

References to website URLs were accurate at the time of writing. Neither the author nor Chemeketa Press is responsible for URLs that have changed or expired since the manuscript was prepared.

Printed in the United States of America.

Land Acknowledgment
Chemeketa Press is located on the land of the Kalapuya, who today are represented by the Confederated Tribes of the Grand Ronde and the Confederated Tribes of the Siletz Indians, whose relationship with this land continues to this day. We offer gratitude for the land itself, for those who have stewarded it for generations, and for the opportunity to study, learn, work, and be in community on this land. We acknowledge that our College's history, like many others, is fundamentally tied to the first colonial developments in the Willamette Valley in Oregon. Finally, we respectfully acknowledge and honor past, present, and future Indigenous students of Chemeketa Community College.

Contents

1 The Scientific Method: What Geologists See

Purpose

❑ An introduction to surface processes and the scientific method.

Materials

❑ Images in the manual or powerpoint images presented by your instructor.

Part 1

Answer the following questions about figure 1.1.

Figure 1.1. Columbia River.

1. Describe the landscape in figure 1.1. List at least five things you see in the picture. If you have taken a geology class before, do not use geologic terms. Describe it as you would to someone who is completely clueless about geology.

2. Suggest an origin for the stair step pattern in the hills across the river. Use your observations from the previous question to support your idea.

3. If you could explore the area further or run any experiment you wish, what would you do or look for to test your hypothesis?

Part 2

Find a partner or two (no more than 3 to a group). Answer the following questions regarding figure 1.2.

Figure 1.2. Little Blitzen Gorge.

4. Describe the landscape in figure 1.2. List at least five things you see in the picture. If you have taken a geology class before, do not use geologic terms. Describe it as you would to someone who is completely clueless about geology.

5. Suggest an origin for the valley in the image. Use your observations from the previous question to support your idea.

6. Was it easier or more difficult to make your observations and hypothesis in a group? What are the advantages and disadvantages to working in groups?

Part 3

Get into a larger group (6–12 people). Answer the following questions regarding figure 1.3 together:

Figure 1.3. East Spring Earthflow.

7. Describe the landscape in figure 1.3. List at least five things you see in the picture. If you have taken a geology class before, do not use geologic terms. Describe it as you would to someone who is completely clueless about geology. Ignore the line drawn in the image for now.

8. Notice the area between the lines. How do you think this formed? Use your observations to support your idea. Work together and come up with one hypothesis as a group.

9. If you could travel to this location, what would you look for to further support your hypothesis?

Part 4

Answer the following questions regarding figure 1.4.

Figure 1.4. Oregon Desert.

10. Describe the landscape in figure 1.4. List at least five things you see in the picture. If you have taken a geology class before, do not use geologic terms. Describe it as you would to someone who is completely clueless about geology.

11. Which way do you think the wind blows on a regular basis? Why?

12. Notice the white areas throughout the photo. If these contain salt deposits, what does that say about the climate in this area (besides what can be seen in the image)?

Name: _____ Lab Time: _____ Due:_____

2 | Introduction to Google Earth Pro

Purpose

In this lab, you will learn how to use Google Earth Pro, and how to extract basic information about topography, landscapes, rock formations, and geologic hazards.

Part 1: Introduction and Setup

Google Earth Pro[1] is a software application you can download for free and that connects to a large Google database and interactively serves back to your computer images of portions of our Earth. The images are a stitched-together, complex combination of air-photos, satellite images and surface topography models, at different resolutions. There are different layers with different types of information that you can turn off and on to customize the information you want displayed in the image. Google Earth has traits similar to traditional maps (an underlying projection with position by latitude and longitude or UTM, a scale bar, representation of geographic features such as roads, and political boundaries). It also offers the ability to view three-dimensional features. If you are interested a fuller and more technical description of Google Earth, one can be found on Wikipedia[2].

You will learn how to navigate in Google Earth to view a variety of geologic surface features. The several labs will require these Google Earth skills:

- ❑ Input latitude and longitude and get to a location of interest
- ❑ Zoom in and zoom out
- ❑ Learn to turn off and on different layers
- ❑ Control oblique view
- ❑ Save image/screenshot and drop into word document
- ❑ Extract the position and elevation of a point
- ❑ Use the ruler tool to measure distance along a straight line or a curving path
- ❑ Use trace tool to identify or highlight a feature
- ❑ Make and save a path
- ❑ Make and save a placemark (pin)

---

1   A web-based Google Earth version also exists but lacks many of the functions needed to complete the Google Earth lab exercises for this course. The mobile version lacks even more. Please don't use these to complete your labs.

2   http://en.wikipedia.org/wiki/Google_Earth

- ❑ Create a topographical profile on land
- ❑ Create a topographical profile on the sea floor
- ❑ Use historical satellite imagery

Before you get started, open the user guide link in this footnote[3] and read over the sections the Essential Google Earth Pro Functions listed here:

- ❑ The Google Earth Pro User Interface
- ❑ The Google Earth Pro Toolbar
- ❑ Navigation
- ❑ The Google Earth Pro Status Bar
- ❑ Searching
- ❑ Working with Places
- ❑ Creating a New Placemark
- ❑ Drawing Paths and Polygons

> Don't stress if you didn't understand everything that you just read. Just having a basic understanding of these functions will make your work easier.

Start Here

Work through the following tasks to complete this lab. Please read instructions carefully and ask questions if you need assistance.

1. First, download Google Earth Pro (if you do not already have it). Google "Download Google Earth Pro," then follow the prompts. This works best on a desktop or laptop. Google Earth Pro cannot be downloaded onto a Chromebook, iPad, or device with limited processing capacity.
2. Go to the 'View' option at the top of the screen. Make sure all these options are checked (figure 2.1):
    a. Toolbar
    b. Sidebar
    c. Status bar
    d. Scale legend

Figure 2.1. Google Earth Pro Required View Options

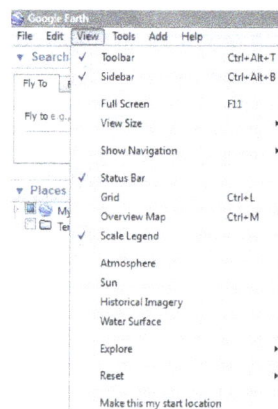

---

3   https://serc.carleton.edu/NAGTWorkshops/teaching_methods/google_earth/UserGuide.html

3. Declutter the map: Uncheck most options under "Layers," including Places, Photos, and Labels. Keep Borders and Terrain boxes checked. Uncheck all other options (figure 2.2).

Figure 2.2. Google Earth Pro Layers Options

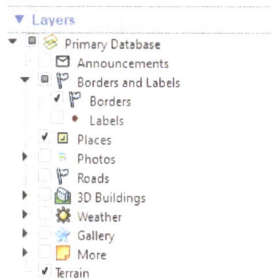

4. Make sure that the units are displayed in the units we've discussed. Go to Tools→Options→3D view tab.
   a. In the "Show Lat/Long" section, select "Degrees, Minutes, Seconds"
   b. In the "Units of Measurement" section, select "Feet, Miles". Click "OK".
5. Finally spend some time familiarizing yourself with how to move and adjust the map view, using both the navigational controls in the upper right corner (figure 2.3) moving the map, zooming in and out, around and using the navigational controls in the upper right.

Figure 2.3. Google Earth Pro Navigational Controls

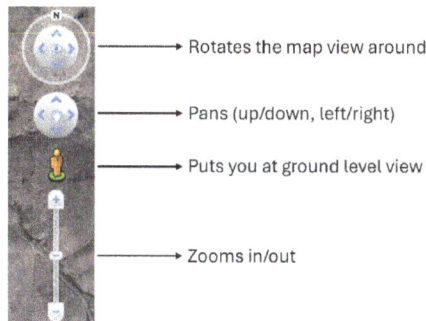

Rotates the map view around

Pans (up/down, left/right)

Puts you at ground level view

Zooms in/out

   a. Zoom in and out with your mouse wheel, mouse pad, or zoom slide bar.
   b. Pro tip: the letter "R" is your friend! If things get weird, just click on the map and type the letter "R". This will make everything right (meaning: north is up and you're looking at the land in map [overhead] view).

Part 2: Using Google Earth Pro

Follow the instructions for each activity and answer each question. Follow the instructions of your instructor to turn in any screenshots or digital images.

Latitude, Longitude, and Elevation of Your Home

Latitude, longitude, elevation and a few other attributes of a place on Earth are shown at the bottom right of your screen and the values change as you move your cursor over the map area (figure 2.4).

Figure 2.4. Latitude, Longitude, and Elevation Display in Google Earth

1. "Fly to" your home by typing in the address in the search bar. Put the cursor directly over your home and zoom in as far can (but not so far that it becomes too pixelated).

1. What is the latitude and longitude of your home? Use the correct format.

2. What is the elevation of your home (in feet)?

3. What is the imagery date of the location of your home?

Measuring Distances

You can measure distances in Google Earth with different tools, but the simplest is the Ruler.

1. While you're looking at your home, measure the width of the street you live on. Select the Ruler tool and a Ruler box will appear. Make sure that feet are selected. Click first on one side of the street, then directly across to the other side. A yellow line will appear and the length of the line you made will show up in the Ruler box.

4. What is the width (length measurement) of your street in feet?

5. What is the width (length measurement) of your street in another unit (your choice!)

6. What is the straight-line distance from your home to Chemeketa Community College (Salem Campus) in miles?

Topography of the US

Topography is the shape of the land surface, which is determined by the distribution of elevation across an area.

1. Zoom out to where most of the contiguous US is shown (lower 48 states only, sorry Hawaii and Alaska!)
2. Move your cursor over the display/map part and observe how the elevation changes as you move over various parts of the contiguous United States. Note how it is related to features like mountains, plains, river valleys, coasts, etc.
3. Look at figure 2.5. Think about how topography/elevation is correlated with the different US geologic provinces. Spend some time doing this.

Figure 2.5. US Geologic Provinces

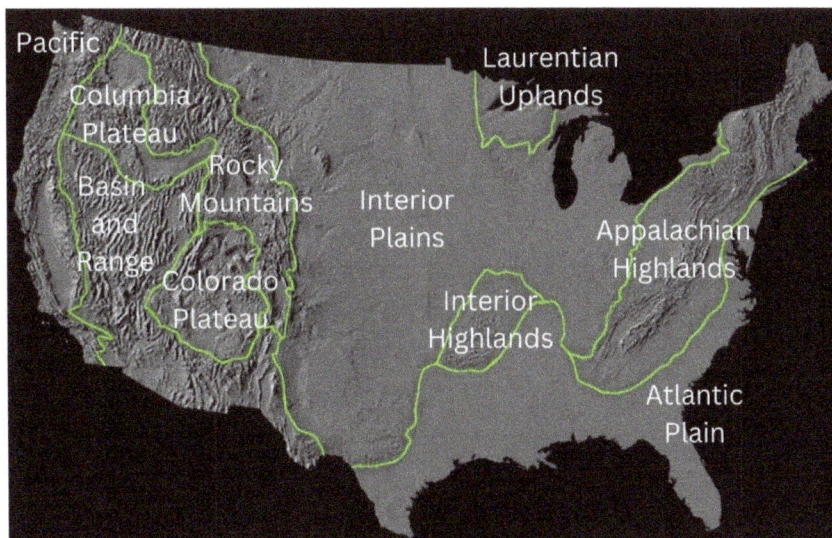

4. Move your cursor around and search for the lowest elevation on land that is **not** adjacent to an ocean.

7. What is it (in feet) and where is it? (indicate general part of specific state and give the elevation). What US Geologic Province is it in?

5. Move your mouse around and search for the highest elevation on land.

8. What is it (in feet) and where is it? (indicate general part of specific state and give the elevation). What US Geologic Province is it in?

Water Depth

Like elevation, water depth can be determined by simply moving your cursor around a body of water and observing the value displayed in the lower right corner. The depth values will be negative since the seafloor is below sea level.

1. Move your cursor offshore (Pacific or Atlantic is fine) and observe how the values change with distance from land and over different seafloor features.

9. Describe how the color of the seafloor is related to depth.

2. Find the location of the Titanic Shipwreck (use the search bar). Zoom in as close as you can.

10. What is the latitude, longitude, and water depth of the Titanic's cold, dark resting place? What is the imagery date?

Elevation Profile Across the US

Earth scientists *love* elevation profiles, which are depictions of a two-dimensional cross-sectional view of a landscape. They highlight the shape of the land along a specific line. Examples are shown in figures 2.6 and 2.7.

Figure 2.6. Example of an elevation profile in Google Earth. The red line on the map is the profile line and the graph at the bottom is the elevation profile (also known as a topographic profile). Snake River Canyon at Oregon-Idaho border.

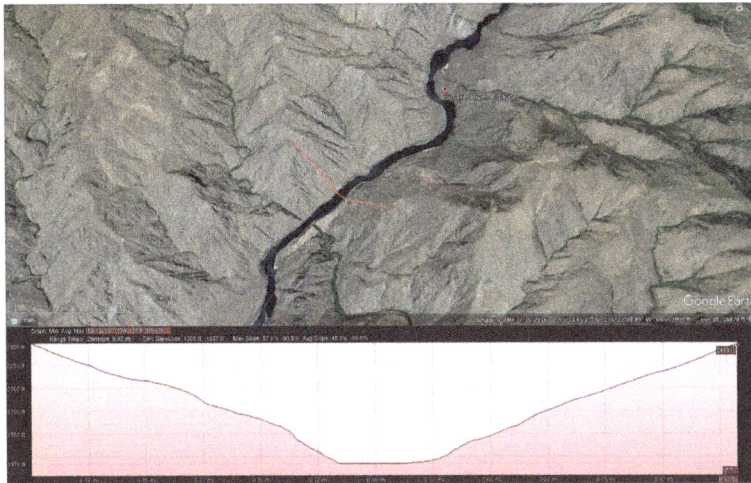

Figure 2.7. Example of an elevation profile in Google Earth. The white line on the map is the profile line and the graph at the bottom is the elevation profile (also known as a topographic profile). Chief Kiawanda Rock (also known as Haystack Rock) off Pacific City, Oregon.

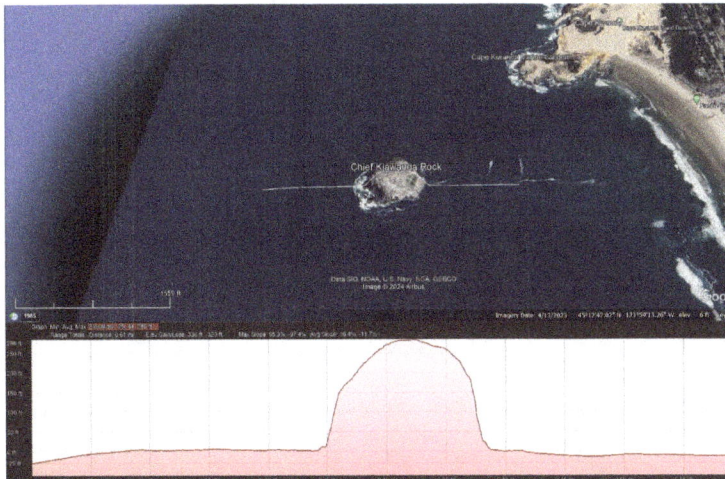

Understanding elevation profiles is a necessary skill for this course. Here you will create a path across the US, from the northern California coast to the Maryland/Virginia coastal area and get a sense of the topography across the US.

1. Select the "Borders and Labels" box in the Layers menu. This will help you see the state lines and a few other things to help you navigate.
2. Select the Add Path tool.
3. Start your path.
   a. First, click on the west coast starting point, any northern California coast beach on the Pacific Ocean.

b. Next, click on the east coast ending point, an Atlantic Ocean beach in the Maryland/Virginia area.

c. Keep your profile line more or less straight. Try to minimize "jumping around" as much as possible. If needed, you can adjust the points when they are highlighted in blue.

4. Give your path a name (e.g., "West to East Coast Profile") and click "OK".

5. Right-click on it in your in the "Places" menu. Click "Show Elevation Profile".

6. Run your mouse along the elevation profile at bottom and notice how the elevation changes along your profile line. A marker arrow will show you the ground location that corresponds with your cursor's position on the elevation profile.

a. Notice that at the top of the profile, some very useful numbers are displayed: elevation minimum, maximum, and average, total length of your profile (distance), slope values, etc.

11. Take a screenshot of the profile image and profile line on the map. Make sure that the full profile line on the map is visible (in addition to your elevation profile). Upload it, **following the directions given by your instructor**.

Hint: Paste your copied image into a Word or Google document. You can upload the document as is, or save the image as an image file by right-clicking on it in the document, then saving it as a picture.

12. Describe your elevation profile. How long is it? How does elevation and topography change as you move from west to east? Be sure to include the relevant provinces, states, and mountain range names in describing this change. This should be at least 4 sentences.

Elevation Profile Across Oregon

This is the same exercise, but using Oregon!

1. Navigate to Oregon.

2. Select "Add Path."

3. Create a relatively straight line across northern Oregon from **west** to **east** that starts right at the coast, runs through Mt. Hood, and ends at the Idaho border.
4. Name the path and click OK.
5. In "Places," right-click on the name of the cross-section you just made and select "Show Elevation Profile." The profile will appear below.
6. Run your mouse along the elevation profile at bottom and make some observations about how the elevation and relief changes.

13. On average, which geomorphic province of Oregon has the lowest elevation (ignoring the area right on the coast)? Highest elevation? Steepest terrain? Flattest terrain? Refer to figure 2.8.

Figure 2.8. Oregon Geomorphic Provinces

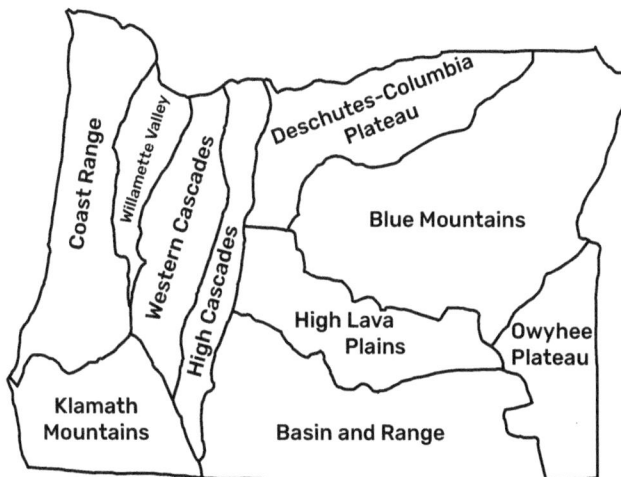

14. What is the total distance of your profile (in miles)?

15. Take a screenshot of the profile image and profile line on the map. Make sure that the full profile line on the map is visible (in addition to your elevation profile). Upload it, **following the directions given by your instructor.**

Placemark a Location

Placemarks are digital "pins" that can be saved to quickly get back to a specific location.

1. Navigate to a place you have a personal attachment to (e.g., Grandma's house; Branson, MO,; Key West, FL; Disneyland)
2. Select the "Add Placemark" tool and move the displayed pin to the exact location.
3. Give it a name and click "OK".

16. Take a screenshot of the profile image and profile line on the map. Make sure that the full profile line on the map is visible (in addition to your elevation profile). Upload it, **following the directions given by your instructor**.

17. Describe it or explain why it is special to you (optional).

Historical Imagery

Satellite images of places are constantly being updated to reflect changes in the land surface due to natural causes (erosion, volcanic eruptions, landslides, fires, etc.) as well as anthropogenic ones (land use changes, construction, etc.). Google Earth displays the most recent images, but stores older ones, too, which can be viewed as far back (generally) as the mid-1980s.

1. While you're still at your "special place" from the previous question, select the "Historical Imagery" tool from the Toolbar.
2. Notice how a slider bar appears. Move the slider around to different dates and observe how the land surface changes over the last few decades.

18. What is the year of the oldest image provided for your location?

19. Describe two specific changes you notice as you move forward in time to the present.

Polygons!

A polygon is a closed, two-dimensional shape that is made up of at least three line segments (e.g., triangles, squares, pentagons, etc.). In Google Earth, a polygon can be used to measure a specific area or simply highlight it.

1. Navigate to a city park near you or one that you liked as a child. Zoom in as far as possible while still being able to view the whole park and center it.
2. Select the "Add Polygon" tool from the Toolbar and click around the margins of the park as carefully as possible. Tthis may take some practice.
3. In the "New Polygon" box that appeared, click the "Style, Color" tab and change the color of the area to a color of your choice (not white). Change the opacity to 50%. (*Hint: if you already closed out of the box, you can get back to it by right clicking on its name in the "Places" menu and then selecting "Properties".*)

20. In the "Measurements" tab, determine the area of the park in square feet and acres.

4. Give it a name and click "OK."
5. Instead of taking a screenshot, this time we will do an image save.
   a. Click the "Save Image" tool in the Toolbar.
   b. Give it a name by clicking on the box that appeared in the upper left corner of the map area.
   c. Click the "Save Image" button that is now beneath the Toolbar.

21. Upload your saved image, **following the directions given by your instructor.**

3 | Sedimentary Rock Identification

Purpose

To identify sedimentary rocks and their textures and apply that knowledge to identifying sedimentary rock environments.

Materials

- ❑ Sedimentary rock kit
- ❑ Dilute HCl
- ❑ Hand lens
- ❑ Glass plates
- ❑ Pennies
- ❑ Nails

Part 1: Sedimentary Rock Properties

Answer the questions about sedimentary rock identification.

1. **Station 1:** The rocks before you are conglomerate and breccia. One of these samples was created in a river environment, and the other was created in a landslide environment. Which sample is conglomerate and which sample is breccia? Support your answers using observations about the rock.

2. **Station 2:** The rocks here are several types of sandstone. What can you see within these samples that caused geologists to place sandstones into different categories? Use more than color to explain your answer.

3. **Station 3:** The rocks at this station are clastic, chemical, and organic. Which sample belongs to which category? Support your answers using observations about the rock.

4. **Station 4:** The samples before you are all monomineralic, meaning that they are chemical sedimentary rocks made from a single mineral. You also have several mineral identification implements before you. Describe how each mineral sample is different by using these implements.

Part 2: Sedimentary Rock Identification

You will be examining and identifying rocks in this part of the lab.

Instructions

1. Use the criteria in table 3.1 to explore rock characteristics by category.
2. Use table 3.2 to identify rock names using observed characteristics.
3. Fill in table 3.3.

Table 3.1. Sedimentary Rock Characteristics

Category	Characteristic
Texture	Is it clastic (weathered pieces cemented together) or crystalline (interlocking crystals)? All detrital rocks are clastic, but chemical rocks can be clastic or crystalline.
Grain Size	How large are the grains? Your options are clay, silt, sand, and gravel.
Rounding	Are the clasts (the pieces of rock found in the sample) rounded like a marble, angular like gravel, or subrounded like Nerds candy? This identifier is only necessary in clastic rocks.
Sorting	Are the clasts well sorted (all the same size in the rock), or are they poorly sorted (with many different clast sizes in the rock)?
Composition	What minerals make up the rock? Some rocks are not composed of one mineral, but may be composed of several rock fragments and do not need a mineral composition.
Environment	Which environment did this rock form? This is determined based on the composition, texture, grain size, rounding, and sorting. Use the following list: • Arid Environment • Beach • Cave • Deep Ocean • Desert Lake • Lake • Landslide • River • Sand Dunes • Shallow Ocean • Swamp • Trench • Tropical Beach

Table 3.2. Sedimentary Classification

Detrital Rocks			Chemical, Biochemical, and Organic Rocks		
Clastic Texture	Grain Diameter	Rock Name	Composition	Texture	Rock Name
Gravel: Round Pieces	> 2mm	Conglomerate	Calcite	Crystalline	Crystalline Limestone
Gravel: Angular Pieces	> 2mm	Breccia	Calcite	Clastic	Oolitic Limestone
Sand: Quartz	.06mm–2mm Visible	Quartz Sandstone	Calcite	Crystalline	Travertine
Sand: Quartz and Feldspar	.06mm–2mm Visible	Arkose	Calcite	Clastic: microscopic shells	Chalk
Sand: Rock Fragments	.06mm–2mm Visible	Graywacke	Calcite	Clastic: shells	Fossiliferous Limestone
Silt	.004mm–.06mm Not visible	Shale	Calcite	Clastic: entirely shells	Coquina
Clay	<.004mm Not visible	Claystone	Quartz	Very fine	Chert or Flint
			Gypsum	Crystalline	Rock Gypsum
			Halite	Crystalline	Rock Salt
			Organic Material	Very fine	Coal
			Dolomite	Crystalline	Dolostone

Table 3.3. Sedimentary Rock Classification Data

ID#	Texture	Grain Size	Rounding	Sorting	Composition	Rock Name	Environment

4 **Topographic Map**

Purpose

❑ To learn how to use topographic maps.

Materials

❑ Salem West Quadrangle map
❑ Mt. Hood South Quadrangle map
❑ Ruler with inches and centimeters
❑ Scratch paper
❑ Calculator
❑ String

Instructions

1. Using the provided maps, answer the following questions.

Part 1: West Salem Quadrangle

1. What is the contour interval of the map? _____ ft.

2. What is the fractional (ratio) scale of the map? _____

     a. 1 inch on the map equals _____ inches on the ground.
     b. 1 centimeter on the map equals _____ centimeters on the ground.
     c. 1 inch on the map equals _____ feet on the ground.
     d. 1 inch on the map equals _____ miles on the ground.

3. How many minutes of latitude does the entire map represent? _____
Minutes of longitude? _____

4. The upper right hand corner of the map is labeled "7.5 minute series (topographic)." What does a 7.5 minute series map mean?

5. What is the elevation of Wallace Hill? _____

6. What latitude and longitude is the Salem Gun Club? _____

    In which direction does Croisan Creek flow? _____
    What evidence do you have to support this?

7. Measure and note the direct distance from the water tank on Illahe Hill to top of Plank Hill in centimeters.
    Map distance = _____ cm.

    Use the fractional scale to calculate the true distance in kilometers. **Show your work below.**

    True distance = _____ km.

8. Calculate the gradient (slope) of Willamette River in feet/mile and as a unitless decimal (or a percent). Show your work below. Follow the river with your finger to find specific elevations:

    ❏ Gradient = _____ ft/mi.

    ❏ Unitless = _____

9. Which side of the Eola Hills is the steepest? _____
    Explain how you determined this:

10. What is the elevation of the Salem Armory? _____

11. What is the highest elevation of the Willamette River on this map? _____

Part 2: Mount Hood South Quadrangle

12. What is the contour interval of the map? _____ ft.

13. Draw a replica of one of the graphic (bar) scales below.

14. What is the elevation of the highest point on the map? _____

15. Using the graphic (bar) scale, how far is it from Government Camp to Alpine
    Campground in miles. _____
    True distance = _____ miles

16. What elevation is Umbrella Falls? _____
    Sahale Falls? _____

17. Calculate the gradient of the East Fork Hood River using the elevations measured from question 16. **Show your work below.**

❏ Gradient = _____ ft/mi.

❏ Unitless = _____

18. What is the elevation of the highest point of Reid Glacier? _____

19. What is located at: SE¼ SE¼ Sec 28 T3S R9E?

20. What is the magnetic declination in this area? _____
What does this mean?

21. On the graphs in figures 4.1 and 4.2, draw a topographic profile between the Bennett Pass and Barlow Pass bench marks (**Neatness and accuracy count!**).

Figure 4.1. 3000' to 5000' Elevation Topographic Profile

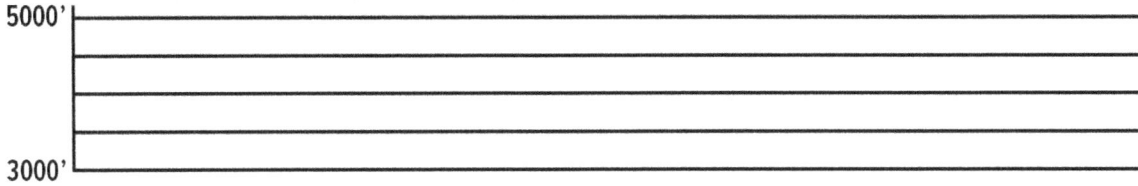

Figure 4.2. 3500' to 4500' Elevation Topographic Profile

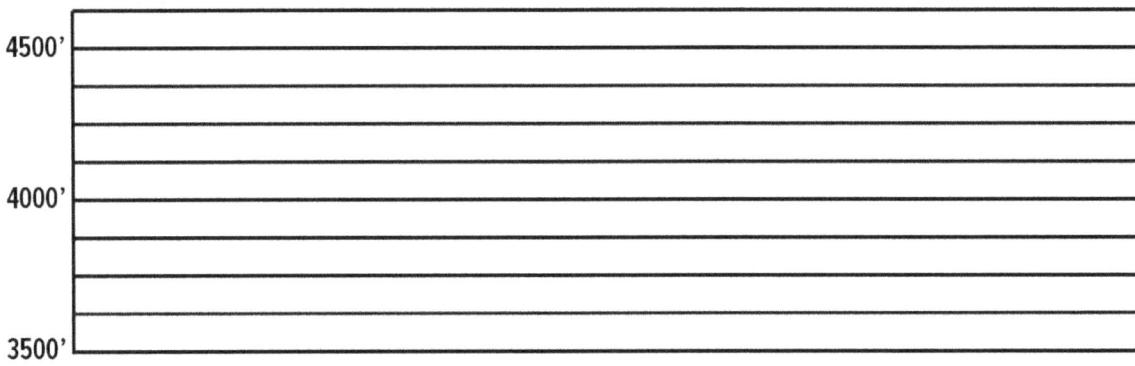

22. Why are the graphs different? What did you do with the second graph that you did not do to the first?

5 | Calculating Stream Discharge

Purpose

In physics, the study of fluid flow has many everyday applications; studying in-stream flow, the movement of groundwater, the flight of airplanes through the air, etc. Because the motion of fluids is so important in everyday life, fluids have been widely studied. In our case, we will be looking at the dynamics of stream flow relating to velocity and discharge and the forces needed to move sediment (rocks and sediment) downstream along a stream bed. You will be both measuring and calculating velocity using Manning's equation.

Materials

- ❑ Water shoes and clothes that can get wet
- ❑ Pencil
- ❑ 50ft Measuring Tape
- ❑ Meter Stick and standard ruler
- ❑ Flagging

- ❑ Flowmeter (orange, marshmallow, plastic ball filled halfway with water, tennis ball, etc.)
- ❑ Calculator (preferably a graphing or scientific calculator)

Part 1: Setup

1. Describe the area of stream you are surveying. What is the vegetation like? How is the weather? Is the bottom of the channel rough? Etc.

2. Measure the width of the stream section you have decided to study and enter the measurement in table 5.1. *Hint: have someone stay on the other bank with the measuring tape stretched across the stream for the rest of the measurements. This will make things easier later.*

3. Divide the stream into three sections by tying flagging tape to the measuring tape as in figure 5.1.

Figure 5.1. Separate Your Stream into Three Sections

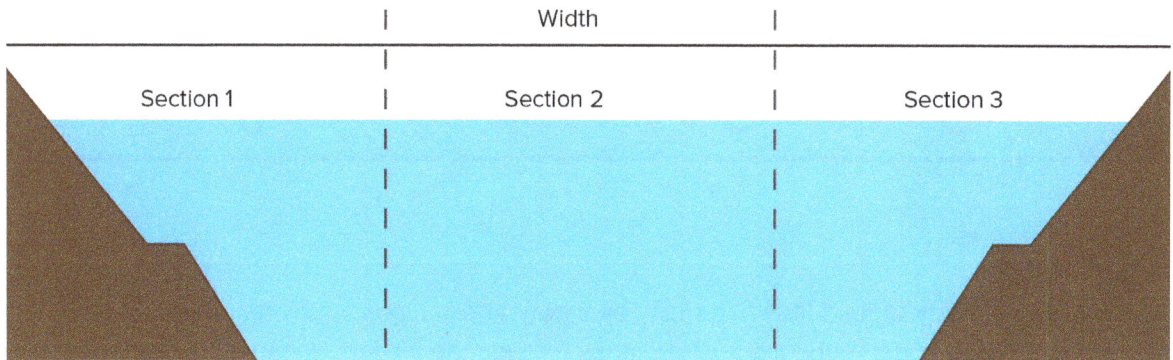

4. Measure and record the depths of the stream at the center of each section. At the same time, take three velocity readings with the flowmeter where you measure depth. Enter your measurements in table 5.1.

5. Once all measurements have been entered in table 5.1, measure the bed load, or the size of the rocks on the bottom of the channel. To do this, have the person coming back from the other side of the stream walk across the channel taking a random sample of 10 different clasts. Measure the longest axis of each sample in cm and record it in table 5.2.

Part 2: Data

Fill out the table 5.1 with the measurements you obtained in the field.

Table 5.1. Stream Parameters

Stream Width			
	Section 1	Section 2	Section 3
Stream Depth			
Stream Velocity Trial 1			
Stream Velocity Trial 2			
Stream Velocity Trial 3			
Average Stream Velocity			

Overall Average Stream Velocity: _____

Table 5.2. Rock Sample Measurements

Rock Sample	Longest Axis Length (cm)	Notes About Sample
1		
2		
3		
4		
5		
6		
7		
8		
9		
10		

Part 3: Calculations

Fill out table 5.3 using the data obtained in the field. **Show your work and maintain your units.**

Table 5.3. Calculations

Stream Area: A = w • d	
Wetted Perimeter: P = w + 2d	
Hydraulic Radius: R = A/P	
*Manning's Equation (velocity): $V = (1.49/n)\ R^{2/3}S^{1/2}$	(n = 0.040, Measure slope, S, in ft/ft, or obtain it from instructor)
Discharge calculated with measured velocity: Q = V • A	
Discharge calculated with Manning's velocity: Q = V • A	

*Manning's equation is an empirical equation that applies to uniform flow in open channels and is a function of the channel velocity, flow area, and channel slope. Manning's n is a coefficient which represents the roughness or friction applied to the flow by the channel

5

Part 4: Questions

1. Sketch the cross-section of your stream channel. It should resemble a graph where the x-axis is width and y-axis is depth of your stream section, like in figure 5.1.

2. How does your flow meter velocity and Manning's equation velocity compare at your site? Which method do you think gives you the most accurate velocity measurement?

3. Why might your calculated velocity be wrong? Do not simply say your measurements or calculations are wrong. Assume you measured everything right and calculated everything right. Why might your measured velocity not reflect the stream's actual velocity for your section?

4. What part of the stream has the highest and the lowest velocities (vertically and horizontally)? Where would you expect the greatest velocities?

5. Do your discharge calculations differ? Should they? Should they if they are the exact same spot? If your discharge calculations differ, why?

5

6. Estimate the size of particles that could be lifted by the velocity recorded in the field using the Hjulström diagram (figure 5.2). What is the velocity needed to lift the largest rocks you measured in the stream and how do they differ from the rocks that can currently be moved by the flow?

Figure 5.2. Hjulström Diagram

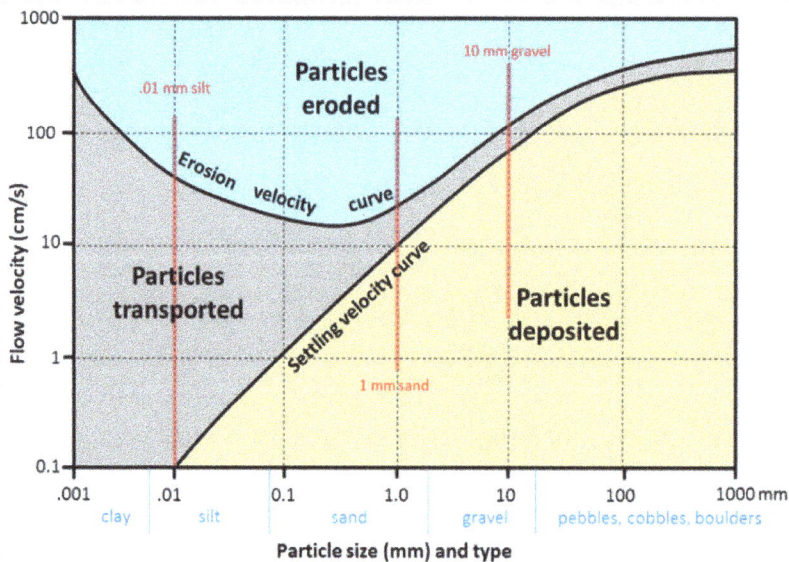

5B Calculating Stream Discharge (alternate)

Purpose

In this lab you are going to calculate the discharge of a small river channel that you make out of household items. Finding the discharge of rivers is important for hydrogeologists because it can be used to estimate the usage of river water over a year. This can be paramount to an area because the flow of a river can be monitored for flood and drought conditions. What you are going to do for this lab is change the characteristics of a handmade river channel to see what will happen to the discharge of the channel as the parameters change.

Materials

- Tin foil
- Duct tape
- Ruler
- Water and pitcher or large bowl

- Small float (i.e., a tiny cracker or a small leaf; anything small enough to fit into the channel that floats)
- Something waterproof to prop up your channel

Part 1: Setup and Data

Read all the methods before you start and take pictures as you do the lab.

1. Make a 24-inch length of tinfoil and fold it in half lengthwise twice. Curve the tinfoil lengthwise and make a half-pipe.
2. Take two pieces of duct tape, about 3 to 4 inches in length, and wrap them over the top of your half-pipe about 12–15 inches apart to hold the curve of the pipe in place (figure 5.3).

Figure 5.3. Making a Half-pipe with Tin Foil and Duct Tape

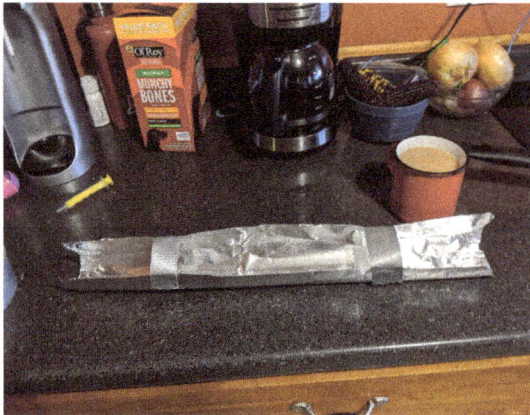

Measure and record the Width of your half-pipe: _____

3. Take your half-pipe outside and lay it on the ground with a very slight incline. Hold the ruler in the half-pipe toward the bottom, past the second piece of duct tape, to measure the water depth as it flows through the half-pipe.

4. Pour water into the channel. As the water runs through the channel, simultaneously:
   a. Measure the depth of the water.
   b. Put the float in the stream of water and measure the time it takes to flow the distance between the two pieces of duct tape.
   c. This might take a few times to get right. What you are trying to do is measure the velocity of the water using the speed of the cracker. Once you get the swing of things, fill out table 5.4.

Table 5.4. Data for Stream Discharge Activity

Test Number	Water Depth	Distance the Float Moved	Time of Float Motion	Velocity of water (Distance/Time)
Example: Test 1	0.5 in	15 in	1.9 s	15 in/1.9 s = 7.9 in/s

5. Calculate the following measurements.

   Average Depth: _____

   Average Velocity: _____

6. Calculate the area of the stream.

   (Average Depth x Width) Area: _____

7. Calculate the discharge of the stream.

   (Area x Average Velocity) Discharge: _____

8. Repeat methods 4–7, but change the parameters of the stream by increasing the slope. After completing the methods, fill out table 5.5:

   Table 5.5. Data for Stream Discharge Activity with Steeper Slope

Test Number	Water Depth	Distance the Float Moved	Time of Float Motion	Velocity of water (Distance/Time)
Example: Test 1	0.5 in	15 in	1.9 s	15 in/1.9 s = 7.9 in/s

9. Calculate the following measurements.

   Average Depth: _____

   Average Velocity: _____

10. Calculate the area of the stream.

(Average Depth x Width) Area: _____

11. Calculate the discharge of the stream.

(Area x Average Velocity) Discharge: _____

12. Once more, repeat methods 4–7, but change the parameters of the stream by increasing the slope and decrease the width of the channel. After completing the methods, fill out table 5.6:

Table 5.6. Data for Stream Discharge Activity with Steeper Slope and Narrower Channel

Test Number	Water Depth	Distance the Float Moved	Time of Float Motion	Velocity of water (Distance/Time)
Example: Test 1	0.5 in	15 in	1.9 s	15 in/1.9 s = 7.9 in/s

13. Calculate the following measurements.

Average Depth: _____

Average Velocity: _____

14. Calculate the area of the stream.

(Average Depth x Width) Area: _____

15. Calculate the discharge of the stream.

(Area x Average Velocity) Discharge: _____

Part 2: Questions

Answer the questions based on your answers from the activity. Make sure that you have taken some pictures of you doing the lab and your channel and make sure you have filled out all of the tables and blanks above and save the information if you are doing the lab at home.

1. What happens to the velocity and discharge as you change the slope from lower to higher? Does this change make sense? Why or why not?

2. What happens to velocity and discharge as you change the width of your channel from larger to smaller? Does this change make sense? Why or why not?

3. What do you think would happen to the velocity of the water through the channel if the bottom of the channel becomes more rough? Why?

4. How would turns, or meanders, in a river affect the velocity?

5. How is your channel similar and how is it different from a channel you might find in nature?

6 Rivers and Fluvial Landforms

Purpose

In this lab, you will use Google Earth Pro to study rivers and fluvial landforms of Oregon and other parts of the world.

Part 1: Setup

Follow the directions below before answering the questions. Follow the instructions of your instructor to turn in any screenshots or digital images.

(Note: the first four steps below may be redundant from Lab 2 as your settings have probably not changed, but it is good to double check. Don't overlook the last 2 steps!)

1. Open Google Earth Pro (or download if using a new device; see Lab 2 for instructions).
2. Go to the 'View' option at the top of the screen. Make sure all these options are checked (figure 6.1):
    a. Toolbar
    b. Sidebar
    c. Status bar
    d. Scale legend

Figure 6.1. Google Earth Pro Required View Options

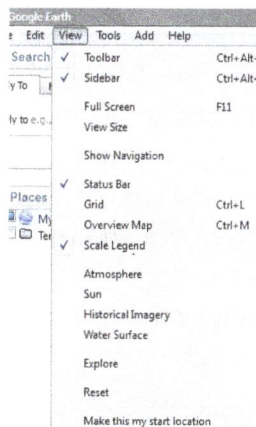

3. Declutter the map: Keep Borders and Labels, Places, and Terrain boxes checked. Uncheck all other options (figure 6.2).

Figure 6.2. Google Earth Pro Layers Options

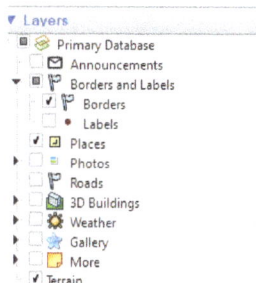

4. Make sure that latitude and longitude are displayed in degrees, minutes, and seconds and that units of measurement are in feet and miles and that the measurement units are English. Go to Tools→Options.
5. Download the Rivers and Fluvial Landforms kmz file (found in Canvas module) to your device.
6. Open the kmz file in Google Earth by clicking File>Open and select the kmz file. This will put placemarks (pins) on your map.

> For several of the questions, you will use the pins placed on your map when you opened the kmz file. Look in the Places menu to ensure that they are there and that the boxes are checked. Click the triangles adjacent to the file name to make sure the pins are shown in list form. This will make it easier to navigate through this part of the lab by double clicking on them.

Part 2: Streams and Valleys of the US

Snake River, Oregon/Idaho Border

1. Go to the Snake River pin.
2. Select the Path tool and click the Measurements tab. Make sure the units for length are miles.
3. Create a path within the river that starts exactly where the pin is and ends at exactly 5.0 miles downstream[1].
   a. Create a path by marking control points spaced closely enough so that the marked path follows the river channel closely.
   b. Zoom in closely and adjust the map as you go so that you do not to click on the rock valley wall—your path should stay 100% in the river.
   c. (*Hints: use the arrow keys on your keyboard to move the map around, your mouse to zoom in and out, and the "R" key to reset perspective of map to straight over-head as needed.*)
4. When your path is **exactly five miles long**, give the path a name and save it. A guide example shown in figure 6.3.

---

1 Elevation values will tell you the direction of downstream.

Figure 6.3. Two-mile path of the Yellowstone River starting below the falls. Note how numerous control points (shown in red) were used so that the path stays in the river instead of running up the valley walls.

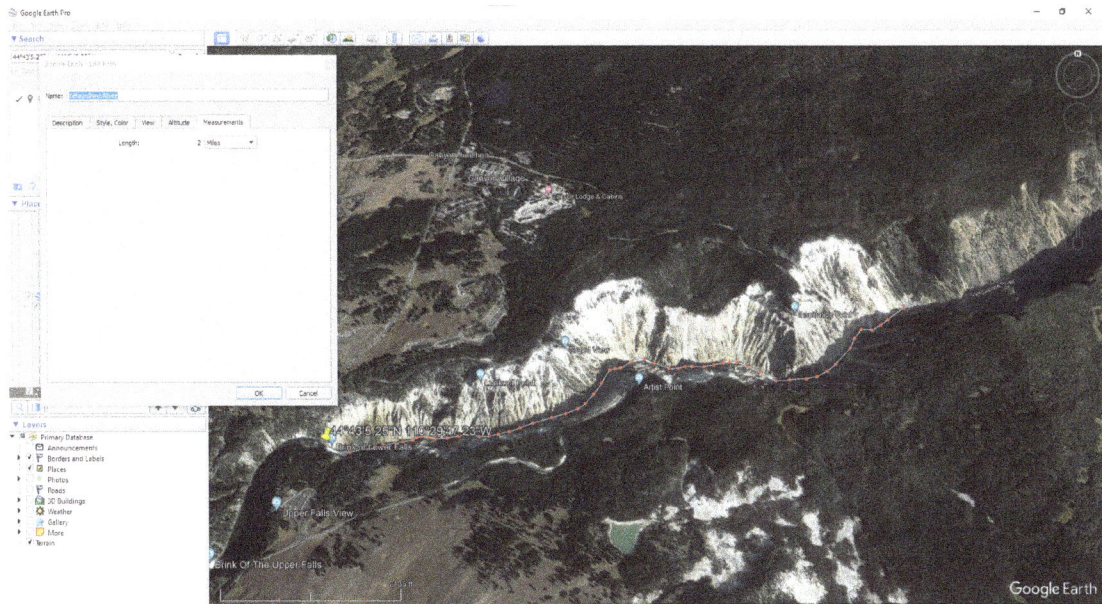

5. Using the Path tool, make an elevation profile[2] and observe how the slope and elevation change along your profile line by moving your cursor along the profile.

1. Carefully observe and briefly describe the appearance of your elevation profile.

2. Take a screenshot that shows the full length of your profile line on the map and the elevation profile. Upload it, **following the directions given by your instructor**.

3. Determine the **gradient** in feet per mile (ft/mi) of this section of the river using the information shown in your elevation profile and the gradient diagram and equation shown in figure 6.4.

---

2   Revisit Intro to Google Earth lab for explanation of how to make an elevation profile.

Figure 6.4. *Gradient* is the slope or steepness of something. It is measured between two points by calculating the change in elevation between the two points and dividing by the distance between them (also known as "rise over run"). Here, you will subtract the low elevation of your profile from your high elevation. Then divide this value by the length of your path. This will be your gradient in ft/mi.

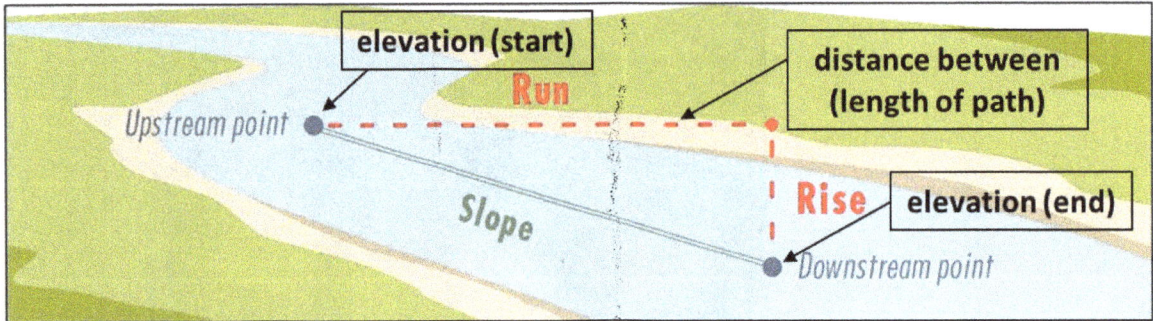

$$\frac{Elevation_{start}(ft)-Elevation_{end}(ft)}{length\ of\ path(mi)}$$

4. Determine the **sinuosity** of this section of the river using sinuosity diagram and equation shown in figure 6.5.

Figure 6.5. *Sinuosity* is a measure of how "curvy" a stream is. It is calculated using the equation shown. The greater the sinuosity value, the curvier the stream. Use the Ruler tool to measure the straight-line distance.

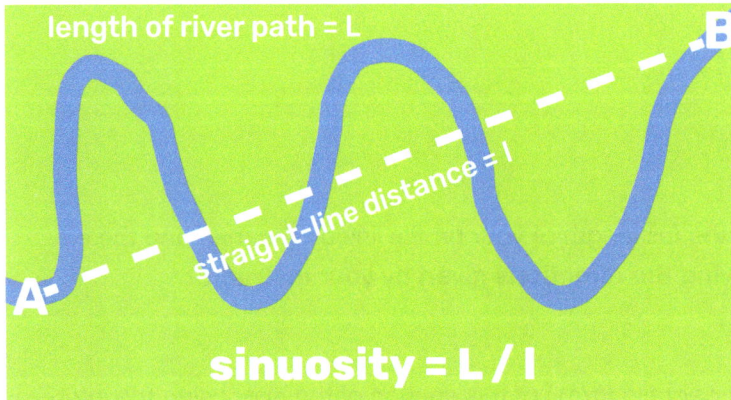

$$\frac{(actual\ length\ of\ river\ path\ =\ l)}{(straight\text{-}line\ distance\ between\ beginning\ and\ ending\ points\ =\ L)}$$

5. Based on your sinuosity calculation, is this section of the river straight, sinuous, or meandering? See table 6.1.

Table 6.1. Sinuosity Types

Type	Sinosity
Straight	<1.1
Sinous	1.1–1.5
Meandering	>1.5

6. Next, we will look at the shape of the river valley in a cross section. Make a path across the river valley that is as close to perpendicular to the flow of river as possible (crossing the first path you drew at a right angle).

   a. Make sure your path covers the entire width of the valley. It should start and end at the ridges on opposite sides of the canyon with a total length of no more than 2 miles. It helps to zoom in significantly when you are doing this.

7. Make an elevation profile for this path and observe the shape of the river valley.

8. Unselect the first Snake River path you made so it disappears from the map.

6. Take a screenshot that shows the cross-valley profile line on the map and the elevation profile. Upload it, **following the directions given by your instructor**.

7. What is the width of the Snake River at this location (in feet)? (*Hint: determine this by clicking and highlighting the segment of your cross section [on the profile] that is the river itself. The value will appear in the "range totals" values above your profile.*)

8. What is this type of stream channel?

9. Explain two reasons for your answer to question 8.

Mississippi River, Louisiana/Mississippi border

1. Navigate to Natchez, Mississippi using the search bar

2. Zoom out and to the south until Baton Rouge, Louisiana, is also visible in the map area. Observe that the Mississippi River that runs between them and constitutes the state border between Mississippi and Louisiana here.

3. In the Layers menu, unselect Borders to make the map less cluttered.
4. Using the same technique as before, draw a path along this length of the river (Natchez to Baton Rouge).
   a. It is critical that your path stays **in** the river and doesn't run up the banks or go across sand bars. You will need to stay zoomed in as you click your control points.
5. Make an elevation profile and observe how the slope and elevation change along the profile line.

10. Take a screenshot that shows the full length of your profile line on the map and the elevation profile. Upload it, **following the directions given by your instructor**.

11. Determine the gradient of the Mississippi River along this segment using the same technique and formula as before.

12. Determine the sinuosity of the Mississippi River along this segment using the same technique and formula as before.

13. Based on your sinuosity calculation, is this section of the river straight, sinuous, or meandering?

6. Notice the abundance of evidence in the landscape that tells us that the Mississippi River has not always flowed along its current course. Navigate your view so that you can see a couple of oxbow lakes and a few meanders.
7. Using the Path tool, highlight[3] (trace) at least two oxbow lakes.
8. Adjust the color[4] of these paths to **light blue** and thickness of the line to 2.0 by clicking on the "Style, Color" tab in the Edit Path box.
9. Using the Path tool, highlight the cut banks/areas of erosion of at least two meanders.
10. Adjust the color of these paths to **red** and thickness of the line to 2.0.

---

3   The Path tool is not just for creating elevation profiles. It can also be used to highlight or draw, in which case, elevation profiles are not necessary (unless otherwise indicated).

4   The style and color of your paths (and polygons) can be changed by going to the "Style, Color" tab of the Edit Path (or Edit Polygon) window. If you've already saved and closed it, you can get back to it by right clicking it in the Places menu and selecting Properties.

11. Using the Path tool, highlight the point bars/areas of deposition of at least two meanders.
12. Adjust the color of these paths to **yellow** and thickness of the line to 2.0.

14. Take a screenshot that shows these three features you have just highlighted in the various places on the map. You can take more than one screenshot in this question if necessary. Upload it, **following the directions given by your instructor**.

13. Find the valley walls on either side of the Mississippi River valley just downstream (south) from Natchez.
    a. The valley walls can be recognized by the predominantly darker color of the wooded terrain of the uplands.
14. Use the Path tool to make a path (cross section) across the valley (perpendicular to water flow), making sure this path extends beyond the edges of the valley walls so that you will see the profile of the entire valley.
    a. *Hint: your total profile length should be between 1.25–3 miles total with the river in the center of it.*
15. Make an elevation profile for this path and observe the shape of the river valley.
16. Unselect the first Mississippi River path you made (so it disappears from the map).

15. Take a screenshot that shows the profile line on the map and the elevation profile. Upload it, **following the directions given by your instructor**.

16. What is the width of the Mississippi River at this location (in miles)? Use the same technique as the previous question.

17. What is this type of stream channel?

18. Explain two reasons for your answer to question 17.

Platte River, Nebraska

1. Navigate to the following coordinates using the search bar: 41°8'11"N, 97°54'36"W
2. Observe the section of the Platte River in Nebraska between the towns of Grand Island and Columbus.

19. Determine the gradient of the Platte River along this segment using the same technique and formula as before.

20. Determine the sinuosity of the Platte River along this segment using the same technique and formula as before.

6

21. Based on your sinuosity calculation, is this section of the river straight, sinuous, or meandering?

3. Near the pinned location (the latitude and longitude you entered), use the Path tool to make a path (cross section) across the river. Start and end in the agricultural areas/fields on either side of the river.
    a. *Hint: your profile should be between 0.5–1 mile in length.*
4. Make an elevation profile for this path and observe the shape of the river valley.

22. Take a screenshot that shows the profile line on the map and the elevation profile. Upload it, **following the directions given by your instructor.**

23. Describe the appearance of your profile.

24. What is this type of stream channel?

25. Explain two reasons for your answer to question 24.

Green River, Washington

1. Navigate to the Green River polygon and observe the shape of the river and landforms surrounding it.

26. What is this type of stream channel?

6

2. In historical imagery, go back in time and view how the stream channel has changed from 1990 until the most recent image. Spend some time observing the changes over the decades.
3. Go to the 1990 image. Using the Path tool, highlight (trace) the main channel of the river within the polygon area. Adjust the color to **dark blue** and the weight of the line to 2.0. Give your path a name that includes the year.
4. Go to the 2002 image. Using the Path tool, highlight the main channel of the river within the polygon area. Adjust the color to **green** and the weight of the line to 2.0. Give your path a name that includes the year.
5. Go to the 2014 image. Using the Path tool, highlight the main channel of the river within the polygon area. Adjust the color to **yellow** and the weight of the line to 2.0. Give your path a name that includes the year.
6. Close out the historical images and observe the various paths as they are overlain on the modern image.

27. Take a screenshot that shows the three highlighted paths. Upload it, **following the directions given by your instructor**.

28. What is going on here? Explain how and why the river has moved over the decades. Explain in a paragraph (4–5 sentences).

29. Your friend tells you that they want to build their dream fishing getaway home on the banks of the Green River in this area. What would you tell them?

Southeastern Colorado

1. Navigate to the following coordinates using the search bar: 37°27'28"N, 103°37'12"W
2. Zoom to an eye alt of about 60 miles and observe the pattern the streams make.

30. What type of drainage pattern is visible?

Mt Rainier, Washington

1. Navigate to this location using the search bar.
2. Observe the pattern the rivers make.

31. What type of drainage pattern is visible?

Part 2: Willamette River Landforms

The headwaters of the Willamette River are considered to be the confluence (meeting point) of the Middle Fork Willamette River and Coast Fork Willamette River, located just south of Springfield, Oregon. The Willamette River mouth is where it meets the Columbia River in north Portland. In this part of the lab, we will look at some of the features along this course.

Dams on the Middle Fork Willamette River

1. Go to the Double Dam pin upstream from the confluence.
2. Note the overall direction of water flow and observe the two dams and the shapes of the reservoirs they have created.

32. What are the names of the two reservoirs that are visible? (*Hint: Adjust your zoom and check both Places and Labels boxes from the Layers menu.*)

3. Zoom closely in on the larger of the two dams and observe the floodgates, the impoundment (reservoir), and the reduced flow side of the dam.
4. Using the Path tool, highlight (trace) the floodgates. Change the color to **dark blue** and adjust the width to 2.0. Name and save this path.
5. Using the Path tool, draw an arrow that indicates the impoundment/reservoir side of the dam (with the arrow pointing in the direction of flow). Change the color to **red** and adjust the width to 2.0. Name and save this path
6. Using the Path tool, draw an arrow that indicates the reduced flow side of the dam (with the arrow pointing in the direction of flow). Change the color to **yellow** and adjust the width to 2.0. Name and save this path.

33. Take a screenshot that shows the three paths you created. Upload it, **following the directions given by your instructor**.

34. Using the Ruler tool, measure the length of the larger dam. *(Hint: Zoom way in so that you can be sure that you are looking at the concrete dam structure specifically, not a parking lot, road, etc.)*

35. Using the same technique, measure the length of the smaller dam downstream.

Confluence of Coast Fork and Middle Fork Willamette Rivers

1. Go to the Confluence pin.
2. Observe the layout of the land and water flow within this area.

36. At this location, the two rivers are behaving like channels of a braided river, as evidenced by numerous mid-channel bars. Why do you think they form here at the confluence?

37. Follow the Willamette River northward to the Harrisburg area. What type of river channel do you observe as a whole?

Harkens Lake

1. Go to the Harkens Lake pin.
2. Observe the shape of the lake and its proximity to the river.

38. What type of lake is this?

Between Albany and Salem

1. Look closely along this stretch of the river and find a meander scar.
2. Using the Path tool, highlight (trace) the meander scar. Change the color to **purple** and adjust the width to 2.0. Name and save this path.

39. Take a screenshot that shows this feature. Upload it, **following the directions given by your instructor**.

40. What river is the largest tributary in this section of the Willamette River?

Willamette Falls

1. Go to the Willamette Falls pin.These are the only waterfalls on the Willamette River and the second largest (by volume) in the US They exist because of a basalt "shelf" in the river at this location.
2. Observe the shape of the falls and the area surrounding them.
3. Check the Places box from the Layers menu (if not already checked).
4. Make an elevation profile starting just upstream from the top of the falls, following the channel (water path) until just past the Willamette Falls Historical site.
   a. You may need to use 3 or 4 control points. Its total distance should be 1500–1600 feet. Observe the slope and elevation change along the profile.

41. What is the height of the falls based on your profile?

42. Take a screenshot that shows the profile line on the map and the elevation profile. Upload it, **following the directions given by your instructor**.

43. This is an area with a significant history of human modifications. Describe two ways (or specific things you see) in which humans have altered this area of the river/falls here.

The Mouth

1. To complete our journey along the Willamette River, find its mouth.

44. What is the name of the point of land where it meets the Columbia River?

7 Teledyne Wah Chang Superfund Site Hazard Simulation

Purpose

This lab is meant to simulate a real world CERCLA superfund site that was remediated by Teledyne Wah Chang and the Environmental Protection Agency. You will use information provided here to create your own remediation plan based on the stake holding parties, scenarios that occurred during the cleanup, and information about remedial plans.

Materials

- ❑ Scratch paper for note taking
- ❑ Calculator

Part 1: Background

Teledyne Wah Chang is a zirconium and non-ferrous metals manufacturing plant that opened in Millersburg, Oregon, adjacent to Albany, and began its current work in 1956. Its main function is to manufacture zirconium metal from zircon sands, but it has also manufactured hafnium sponge, tantalum, and niobium in pilot projects (EPA.gov, 2017). The fabrication of these metals creates different waste products including sludge, wastewater, residues, and gases. These are referred to as lime solids. When the manufacturing plant was first opened, and prior to 1979, the wastes that were created during manufacturing were placed in unlined ponds on the plant's property including the Lower River Solids Pond, Schmidt Lake, Arrowhead Lake, and the V2 pond (EPA.gov, 2017). After 1979, lime solids were stored in ponds that were constructed in an area referred to as the farm site (EPA.gov, 2017).

Teledyne Wah Chang's facilities are located on 225 acres of land that include a 110-acre plot containing the main plant and another 115-acre plot called the farm site, where storage ponds are located. The main plant is broken into different areas, including the extraction area, the fabrication area, and the solid storage ponds (EPA.gov, 2017). The main plant is bound by Interstate 5 to the east and Old Salem Road to the west. About half a mile to the west of the main plant is the Willamette River, and much of the land to the east and the west of the main plant is used for residential, commercial, and agricultural purposes. Part of the facility is in 100-year and 500-year floodplains (EPA.gov, 2017). Underlying the facilities is a rock formation called the Linn Gravel, which is designated as a groundwater source for the Albany area.

For many years, the Environmental Protection Agency (EPA) and the Oregon Department of Environmental Quality (DEQ) had been concerned about radioactive materials in the storage ponds. Their presence was confirmed by the Oregon State Health Division (OSHD) in 1977. In October 1983, the site was placed on the National Priorities List (NPL) under the Comprehensive Environmental Response, Compensation, and Liability Act (CERCLA), also known as Superfund. CERCLA was passed in 1980 in response to problematic hazardous waste practices and management.

Though there was some remediation at Teledyne Wah Change between 1983 and 1988, on December 28, 1989, a Record of Decision (ROD) for an interim Response Action was made to begin cleanup on the sludge ponds and Schmidt Lake (EPA.gov, 2017). An ROD for an interim Response Action is a legal decision that allows the area to begin the process of cleanup before a final decision for remediation is made. Also, during the next several years, public comment could be submitted in regards to the final cleanup plan being implemented.

7 | Part 2: Community Concerns

Your instructor will separate you into six community groups who were stakeholders in the creation of the Teledyne Wah Chang Superfund Site. Once you have separated into your respective groups, your group will receive a description of your position with regard to the cleanup. As a group, you will come up with a list of your priorities for the cleanup. Enter those here.

Community Group: _____

List of Priorities:

Part 3: Initiating Remediation OU1 (Operation Unit)

Now that you have created your community group's list of priorities, take your list and join a new group. Your instructor will separate you into groups that contain at least one representative from each community group. Once you are in your group, you will share your priorities with each other and try to come to a consensus about how you will begin cleaning up the sludge ponds.

To clean up the sludge ponds, you will be removing solid materials through excavation, literally removing the waste material and trucking it to a processing facility. Excavation costs between $150 and $200 per cubic yard removed.

1. What does your group think is the most important first step in remediation? Second step? Third step?

7

2. Did your group choose to leave the plant open during the cleanup operation? Why or why not?

3. Did your group choose to prevent access to the public lakes during the initial cleanup?

4. On your first round of cleanup, you removed 100,000 cubic yards of solids from the sludge ponds and 2,016 cubic yards from Schmidt Lake. Based on the information above, how much money have you spent on cleanup at this point?

Part 4: Initiating Remediation OU2

While cleaning the initial solid wastes in OU1, a full study of the area showed that there was widespread groundwater and surface water contamination of trichloroethane and trichloroethene (Noel Mak, 2017 personal communication). These compounds have been linked to short-term central nervous system depression and long-term abnormalities in the kidneys, liver, and heart. Also, there were some areas tested that showed local groundwater contamination of fluoride and radioactive radium isotopes (Noel Mak, 2017 personal communication). The final decision regarding the extent of contamination of the area called for 15 years of groundwater monitoring and treatment. The costs of placing and operating wells to remediate the area are as follows:

❏ Install single water well and extraction system ($2000)
❏ Extracting and air/liquid phase carbon adsorption stripping ($250/day)

Your team will use the map in figure 7.1 to position wells around the superfund area.

Figure 7.1. Site Map and Grid Overlay

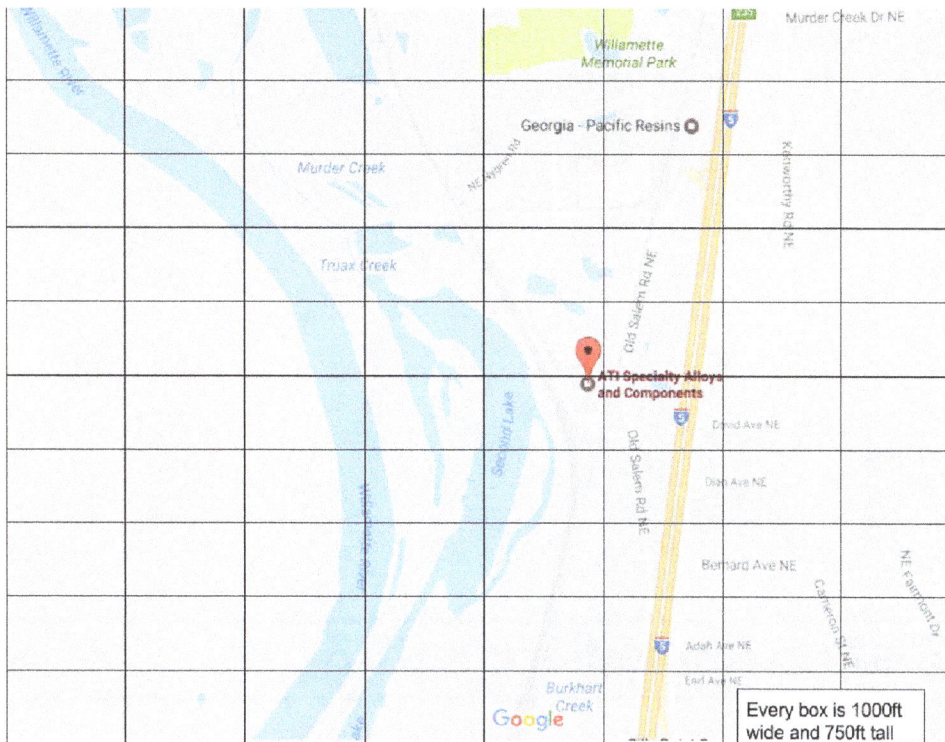

Every box is 1000ft wide and 750ft tall

5. As a team, choose where you would like to put your extraction wells on figure 7.1. Mark them on the map. Once in place, calculate how much your well placements will cost. Show your work.

6. Now that you've placed your wells, calculate the total cost of operation over the 15 years of extraction and treatment. Show your work.

7. With your wells in place, do you think it's necessary to restrict the use of the water bodies near the areas? If you do restrict the area, how will you restrict them and for how long? Things you can restrict include (but are not limited to) access, fishing, purchase of land for private use, farming, groundwater extraction, etc.

Part 5: Initiating Remediation OU3

During the full study of the area, it was also found that the main plant area and an area referred to as the soil amendment area were contaminated with PCBs and zircon sands, which were emitting gamma radiation (figure 7.2).

Figure 7.2. Aerial Photo of Site Area (Oregon Department of Human Services, 2009).

In the human body, gamma radiation causes mutations in DNA, damages cellular mechanisms, and, in large doses, it can kill cells and cause radiation poisoning. In response, another operating unit was created to remove contaminated surface and subsurface soils. Eighty-five cubic yards of material was excavated and removed.

Under Teledyne Wah Chang's natural gas and electric generating plant, wastes were found with high radium-226 concentration. However, the plant was built on a 14-inch-thick cement foundation that effectively blocks gamma radiation, so these wastes were left in place.

8. Using the costs from part 2, how much would it cost to remove 85 cubic yards of soil materials?

7

9. Radium-226 decays into radon gas. Over time, radon exposure can cause lung cancer. In fact, it is the second leading cause of lung cancer after cigarettes. What sort of precautionary measures would your group put into place to prevent future radon exposure to humans in the area?

Part 6: Review

10. Review all of the sections and add up the total costs of your remediation. How much did it cost you to clean up the site?

11. Actual estimated cost of remediation is $112 million dollars and still growing (Noel Mak, 2017 personal communication). Did your group spend more or less on the project than was spent in reality?

7

12. Knowing that several more types of contamination were found after the initial cleanup process, would your group have continued to keep the plant open and in operation like Teledyne Wah Chang did? Why or why not?

13. This simulation was much less complicated than the actual cleanup operation, but it gives you a small glimpse of processes that occur on Superfund sites. Based on your experience today, do you think the government should enforce the cleanup of contaminated sites or should these sites be dealt with privately? Why or why not? **Please write at least a paragraph explaining your position in depth.**

Bibliography

EPA, 1989. Record of Decision, Decision Summary and Responsiveness Summary for Interim Response Action, Teledyne Wah Chang Albany Superfund Site, Operable Unit #1 Sludge Pond Units, Albany, Oregon. Prepared by the United States Environmental Protection Agency. June 10, 1994.

Gearheard, Michael F., and E. P. A. Region. "Second Five-Year Review Report for Teledyne Wah Chang Albany Superfund Site Millersburg, Oregon." Analysis 4 (2003): 1-1.

Oregon Department of Human Services, 2009. Public Health Assessment Final Release: ATI Wah Chang (Formerly Known as Teledyne Wah Chang, Millersburg, Oregon. Prepared under a Cooperative Agreement with the US Department of Health and Human Services and the Agency for Toxic Substances and Disease Registry Division of Health Assessment and Consultation Atlanta, Georgia 30333.

"Search Superfund Site Information." EPA. Environmental Protection Agency, 20 Feb. 2014. Web. 04 Feb. 2017.

8 | Groundwater Mapping

Purpose

In this lab you will create a contour map of a hypothetical water table and track pollution through the groundwater system, using data limited by budget and accessibility of the aquifer.

Part 1: Background

A local college learns that the level of heating oil in one of their underground storage tanks is lower than would be expected. They do not know how long this has been the case. They perform a tightness test on the tank, its lining, and its lines and discover that there is a leak in the system. They need to hire an environmental consulting firm to determine the extent of the leak and how to clean it up.

You are the consulting firm they hire. You will need to stay within your contracted budget while investigating the possible leaks. You will use the map in figure 8.1 on page 70 to plan and evaluate this project.

Part 2: Drilling Wells

First, you need to drill wells near the tank to determine if it actually leaked into the groundwater. Then, if there is a leak, you will need to figure out the extent of the leak by drilling more wells.

For each well, you measure the elevation of the water table. You can also send the water samples to a lab to test for the Total Petroleum Hydrocarbons (TPH) to show the level of contamination (table 8.1).

Table 8.1. Total Petroleum Hydrocarbon (TPH) Values

THP Value (parts per billion)	Interpretation
Less than 200 ppb	Clean
Between 200–20,000 ppb	Contaminated
More than 20,000 ppb	Very contaminated

Figure 8.1. Groundwater Contamination Map

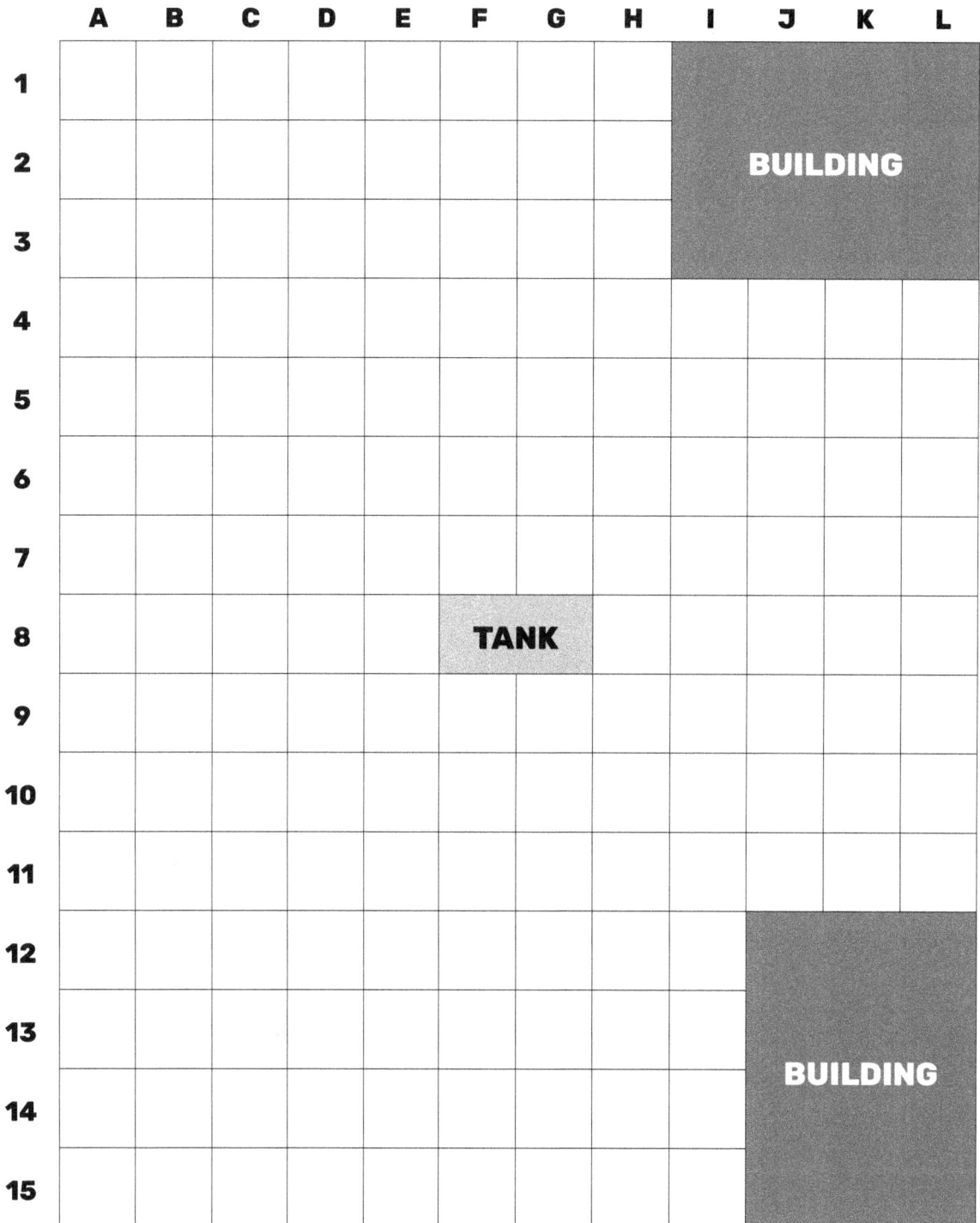

The college hires you for a total budget of $5000. It costs $1000 to drill 4 wells, and this includes measuring the elevation of groundwater. Because you hire the driller for a day, you have to drill 4 wells at a time. The TPH test costs $75 for each sample, and there is no daily limit. Be sure to keep track of your spending (your instructor won't until the lab is graded). You cannot drill on the tank or in the buildings.

Once you determine the locations of the wells you want to drill, ask your instructor for the data. Be sure to specify the location of each well you want to drill, and whether or not you are testing for contamination (and which wells in the group you are testing). The instructor will give you the elevation of the water table (in meters above sea level) for each well, and any requested TPH data.

The TPH test can be done at any time for any well you have drilled. You do not have to request the tests at the same time as the wells are selected, and can come back later to get contamination levels at any time so long as it is within your budget to do so.

Part 3: Making a Map

After determining the extent of the leak (while staying within your budget), you need to prepare a map for the college that hired you on figure 8.1. Your map **must** include:

1. The data provided by your requests:
   a. Water table elevations
   b. Any TPH data for each well location you tested
2. The extent of the leak.
   a. Be sure to "fill in the gaps" so your map shows where you interpret the entire leak to be, not just where you have measured it to be.
3. Arrows showing the direction of movement.
   a. Groundwater flow travels perpendicular across contour lines, so as you create your arrows make sure they cross your contours at right angles.
4. Groundwater contours labeled. Use contour interval of 0.5 meters.
   a. Provide contours for the whole map, not just the areas you drilled.

Part 4: Questions

After creating your map, use it to answer the following questions.

1. Given the extent and location of the contamination, what issues are there with cleaning or addressing the issue?

2. What is your recommendation for solving this contamination problem? What would you suggest to the college that hired you?

3. What information are you missing for a properly informed answer to question 2? What things would be helpful to making the best decision about what to do with the contamination? Consider our discussions on groundwater flow, and what you know and don't know with our current map.

9 Glaciers and Floods

Purpose

For this lab we will be using several maps to explore features created by glaciers as well as features made by the Missoula Floods.

Materials

- ☐ Mowich Lake Quadrangle
- ☐ Drake Crossing Quadrangle
- ☐ Coulee City Quadrangle
- ☐ Appledale Quadrangle

Instructions

Answer the questions using the maps provided. Use the graph paper on page 78 to draw topographic profiles. Your profiles will be graded based on accuracy and neatness, so please take your time completing them.

Part 1: Rivers vs. Glaciers

Maps: Mowich Lake and Drake Crossing

1. Look at the two maps provided. What are the similarities between the two terrains?

2. What are the differences between the two terrains?

3. Using graph paper (p. 78), make a topographic profile going across the Silver Creek canyon. Use a vertical exaggeration of 2×. This means we will expand the vertical scale to twice what it is for the horizontal scale. The horizontal scale is 1:24000 (1 inch = 2000 feet). If we stretch this, it becomes 1:12000 (1 inch = 1000 feet).

4. Make a second profile across the Carbon River Valley (where there is no glacier). Use the same vertical exaggeration.

5. Why are we using a vertical exaggeration?

6. On each profile, mark where the stream is. Be sure it is accurate to the scale of the profile.

   What are the differences in stream locations, relative to the rest of the canyon, for each profile?

   Why are they different?

7. Is there any evidence of glaciers being present within Silver Creek? Why or why not?

Part 2: Mountain Glaciers

Map: Mowich Lake

8. Find Goat Island Rock. Notice how the Carbon Glacier splits around it. In the smaller lobe to the east, what is the current elevation of the terminus?

9. What was the elevation of the glacial terminus (of the Goat Island lobe) in the past? If it is different, this suggests it has retreated. What evidence do you have of this?

10. Explain the formation of the multitude of waterfalls falling into the valley of the Carbon River.

9

11. What is the channel type of the Carbon River? Why is it that type of stream?

12. Why is there a marshy area in the Elysian Fields? Explain in the context of glacial processes and features.

13. Make another topographic profile, this time across the Carbon glacier. Same rules as before. Notice how the contour lines indicate the top of the ice. In addition to the profile, draw a line below the surface of the glacier showing where the bottom of the valley is. You will have to estimate the thickness of the glacier.

14. What could you use to estimate the elevation of the equilibrium line for this glacier? The equilibrium line is the boundary between the accumulation and ablation zones.

Part 3: Missoula Floods

Maps: Appledale and Coulee City

15. Make a topographic profile across Moses Coulee. Use the same rules as before.

16. How is the shape of the coulee (your new profile) different than the canyons of Silver Creek or Carbon River?

17. Along the northern cliff of the coulee there is a gentle dropoff leading into the canyon floor. If this lower terrace were made of very coarse gravel and boulders, how do you think it formed? Explain your reasoning, but keep in mind the coulee is a feature carved by the Missoula Floods.

18. How tall is Dry Falls in feet? How wide are the falls in feet? Include the dropoff not labeled as part of the fall.

19. Besides the dropoff itself, what other evidence is there that water spilled over Dry Falls?

20. Look at a Google Earth or Google Satellite image of Grand Coulee. What evidence is there that Dry Falls moved through the coulee over a period of time and multiple flood events?

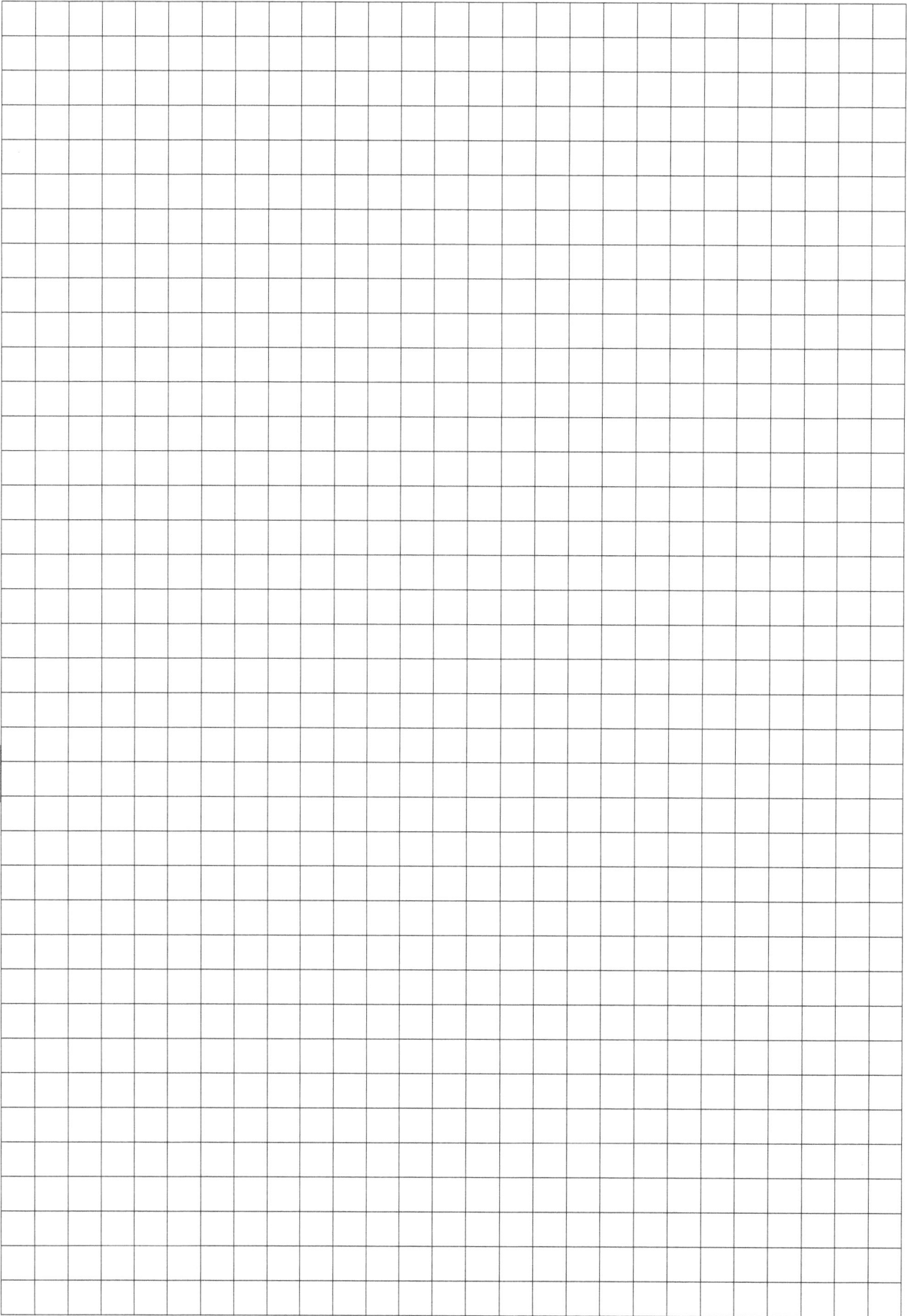

10 Glacial Flow and Retreat

Purpose

This lab will explore the physical motion of glaciers. It also uses real world examples of glaciers in Glacier National Park, Montana, to calculate actual rate of glacial retreat since the Little Ice Age.

Part 1: Glacial Flow

Materials

- Particleboard or other flat surface
- Ruler
- Paper circle of three different colors
- Wax paper
- Piece of tape
- Permanent marker
- Glacial goo (Elmer's Glue, for example)

Methods

1. Set the particle board flat on the table
2. Stretch one batch of glacial goo so it is about 10 cm in length and 5 cm in width and place it on your particle board with the longside aligned with the longside of the board.
3. Place three orange dots in a row across the upper portion of the surface of the glacial goo. Place 3 green dots across the middle of the glacial goo. Place the 3 pink dots across the lower portion of the glacial goo (figure 10.1).

Figure 10.1. Setup of Glacial Goo Activity

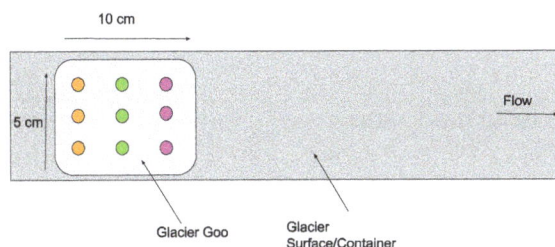

4. Place the sheet of wax paper on top of the glacial goo.
5. Label the location of each placed colored dot on the wax paper with a zero noting the color of the dot.
6. Elevate the short edge of the particle board by about 6 inches and let the glacial goo flow.

7. After 1 minute, lower the board and mark the new location of each colored dot by placing the wax paper back on the glacial goo with the number 1. After minute 2, mark the new location of each dot with a 2 the same way and so on for 5 minutes (unless it reaches the desk surface beforehand).
8. When completed, lower the board so it is flat on the table.
9. Remove the dots on top and carefully lift up the lower portion of the glacial goo without stretching it to find dots that moved under the glacial goo. Try to measure how far back they are from the leading edge of the glacial goo and mark it at the appropriate length on the wax paper sheet.
10. Using one marker at a time, connect the three 0s of one color with a line, then the three 1s with a line,and so on. Do the same for each color of dots.
11. Notice the pattern of the surface flow of the glacial goo.

Using the results of your experiment, answer the following questions.

1. Are all the dots moving in a straight line downhill?

2. Did all three colors of dots move at the same speed downhill?

3. Did the three dots of one color move at the same speed downhill?

4. Which set of colored dots move the fastest downhill?

5. Which one of the 3 pink dots moved fastest downhill?

6. Calculate the speed of the middle dot in each color from its starting point at 0 minute to its location at **three** minutes by measuring the distance between the two dots. (speed = distance / time) The units will be cm/minute.

Speed of middle pink dot: _____

Speed of middle green dot: _____

Speed of middle orange dot: _____

7. Based on your observations and your calculations, what conclusions can you make about the surface flow of a glacier? (where is it moving faster and slower and **why** it is moving faster or slower in those places)

10

8. What would you expect the pattern of the movement of dots to look like on the wax paper if you were to raise the particle board to a much higher elevation?

Part 2: Percent Loss of Glacier in Glacier National Park

Glacier National Park was founded on May 11, 1910, when President Taft signed legislation for the park's creation. It was the tenth national park to be created. The creation of the park intended to protect the area's habitat and topography, but also the many glaciers in the area.

Flash forward more than 100 years and Glacier National Park is losing many of the glaciers that the park was named for. Using geologic records from the Little Ice Age as well as measurements of ice extent from the years 1966, 1998, 2005, and 2015, you will calculate the percent change of glacial area for three of the park's glaciers and answer questions based on how the glacier is changing over time.

Materials

- ❑ Calculator
- ❑ Pencil and Paper

- ❑ Table 10.1. Areas of Glaciers

For each glacier, use the data provided in the data table to calculate how the area of the glacier has changed over time using the percent change formula:

$$|(Final - Original) / Original| * 100 = ?$$

Table 10.1. Areas of Glaciers

Glacier Name	Area LIA (km²)	Area 1966 (km²)	Area 1998 (km²)	Area 2005 (km²)	Area 2015 (km²)	% Decrease from LIA to 2015
Boulder	0.8296	0.2310	0.0488	0.0458	0.0353	96
Grinnell	1.9756	1.0202	0.7159	0.6156	05637	71
Salamander	0.2501	0.2290	0.1817	0.1736	0.1761	30

Data extracted from USGS "Area of the Named Glaciers of Glacier National Park and Flathead National Forest at the Little Ice Age maximum extent, 1966, 1998, 2005 and 2015."

Boulder Glacier

Figure 10.2. Boulder Glacier, circa 1910 (left) and 2007 (right)

Boulder Glacier

9. What was the percent change between the LIA and 1966?

10. What was the percent change between 1966 and 1998?

11. What was the percent change between 1998 and 2005?

12. What was the percent change between 2005 and 2015?

13. What was the percent change between the LIA and 2015?

14. Does calculating the percent change of glacial area between the LIA and 2015 do a good job of showing how the glacier has changed over time? Why or why not?

15. Calculate the rate of retreat of this glacier by using the following equation:

(1966 area – 2015 area) / time of retreat = Rate of Retreat

Based on your rate of retreat, does this glacier still exist? If it does, when will it be completely melted? If it doesn't, when did it disappear?

Grinnell Glacier

Figure 10.3. Grinnell Glacier, circa 1910 (left) and 2016 (right)

16. What was the percent change between the LIA and 1966?

17. What was the percent change between 1966 and 1998?

18. What was the percent change between 1998 and 2005?

19. What was the percent change between 2005 and 2015?

20. What was the percent change between the LIA and 2015?

21. Why does the loss of area with this glacier seem to decrease over time? Should we expect these results?

10

22. Calculate the rate of retreat of this glacier by using the following equation:

(1966 area – 2015 area) / time of retreat = Rate of Retreat

Based on your rate of retreat, does this glacier still exist? If it does, when will it be completely melted? If it doesn't, when did it disappear?

23. Do you think anything could change the speed at which this glacier melts? How could the speed of melting speed up? How could the speed of melting be slowed? (you can use natural or man-made interventions for this question)

Salamander Glacier

Figure 10.4. Grinnell and Salamander Glaciers, 1938 (left) and 2019 (right)

24. What was the percent change between the LIA and 1966?

25. What was the percent change between 1966 and 1998?

26. What was the percent change between 1998 and 2005?

27. What was the percent change between 2005 and 2015?

28. What was the percent change between the LIA and 2015?

29. Why do you think this glacier has not melted as quickly as the other glaciers that we looked at? Are there topographic or geologic differences that could be protecting this glacier from melting?

10

30. Calculate the rate of retreat of this glacier by using the following equation:

(1966 area – 2015 area) / time of retreat = Rate of Retreat

Based on your rate of retreat, does this glacier still exist? If it does, when will it be completely melted? If it doesn't, when did it disappear?

Bibliography

USGS Northern Rocky Mountain Science Center, Repeat Photography Project, https://www.usgs.gov/centers/norock/science/repeat-photography-project, April 6th, 2016. (All related pictures are courtesy of this project)

USGS Northern Rocky Mountain Science Center, Status of Glaciers in Glacier National Park, https://www.usgs.gov/centers/norock/science/status-glaciers-glacier-national-park?qt-science_center_objects=0#qt-science_center_objects, April 6th, 2016.

USGS Northern Rocky Mountain Science Center, Time Series of Glacier Retreat, https://www.usgs.gov/centers/norock/science/time-series-glacier-retreat May 9, 2017.

10

11 | Coasts

Purpose

To use topographic maps and Google Earth to explore the features and processes of coastlines.

Materials

- ❑ Winchester Bay Quadrangle
- ❑ Cape Sebastian Quadrangle
- ❑ Crater Lake National Park topographic
- ❑ Ruler

Instructions

Answer the following questions using the provided maps. Use the graph paper on pages 93–94 to draw topographic profiles.

Part 1: Cape Sebastian Quadrangle

1. What process is responsible for the formation of Hunters Cove (the bay forming behind the headland)? Make a sketch of the cove showing the direction of waves in the area.

2. What type of rock do you think is around Cape Sebastian, and why would it explain the presence of multiple headlands?

3. Why does the beach stick out further behind the labeled "rocks" in Section 7? Draw another sketch showing the waves along this part of the shoreline.

4. Explain the elevated beach behind Crook Point.

11

5. Make a topographic profile across the coastline using the graph paper on pages 93–94. Start your profile from the 861 ft. elevation marker in section 5, T38S R14W, and go west through Hunter's Island. What features of the profile indicate that this area is primarily an emergent coastline? You may select your own vertical exaggeration, but make sure to list it here.

Part 2: Applegate Quadrangle

6. Which way is the longshore transport in this area? What evidence from the map did you use? Be specific, not just "the shape of the shoreline."

7. What about this map suggests a submergent coastline? Why would an area we just described as emergent be submergent as well (both maps are in Oregon)?

8. Make a topographic profile across this shoreline (ask your instructor where). You may select your own vertical exaggeration, but make sure to list it in your profile.

9. How is this profile different from the one you did by Cape Sebastian?

11

10. What type or types of dunes are present along this beach? What evidence (from the map and/or Google Earth) supports your conclusion?

Part 3: Crater Lake

11. How deep is the lake (according to the map)?

12. The lake formed because the rock surrounding the caldera is somewhat impermeable. What evidence from the map suggests that water leaks out, though slowly?

13. Besides your answer to question above (which is based on evidence from the map), what is another way (that wouldn't be seen on the map) that water might be removed from the lake?

11

14. Make a topographic profile across the middle of the lake (from west to east), including Wizard Island. *Note: be sure to catch what the blue contour lines are indicating (it is not elevation above sea level)*. Include the elevation of the lake surface in your profile.

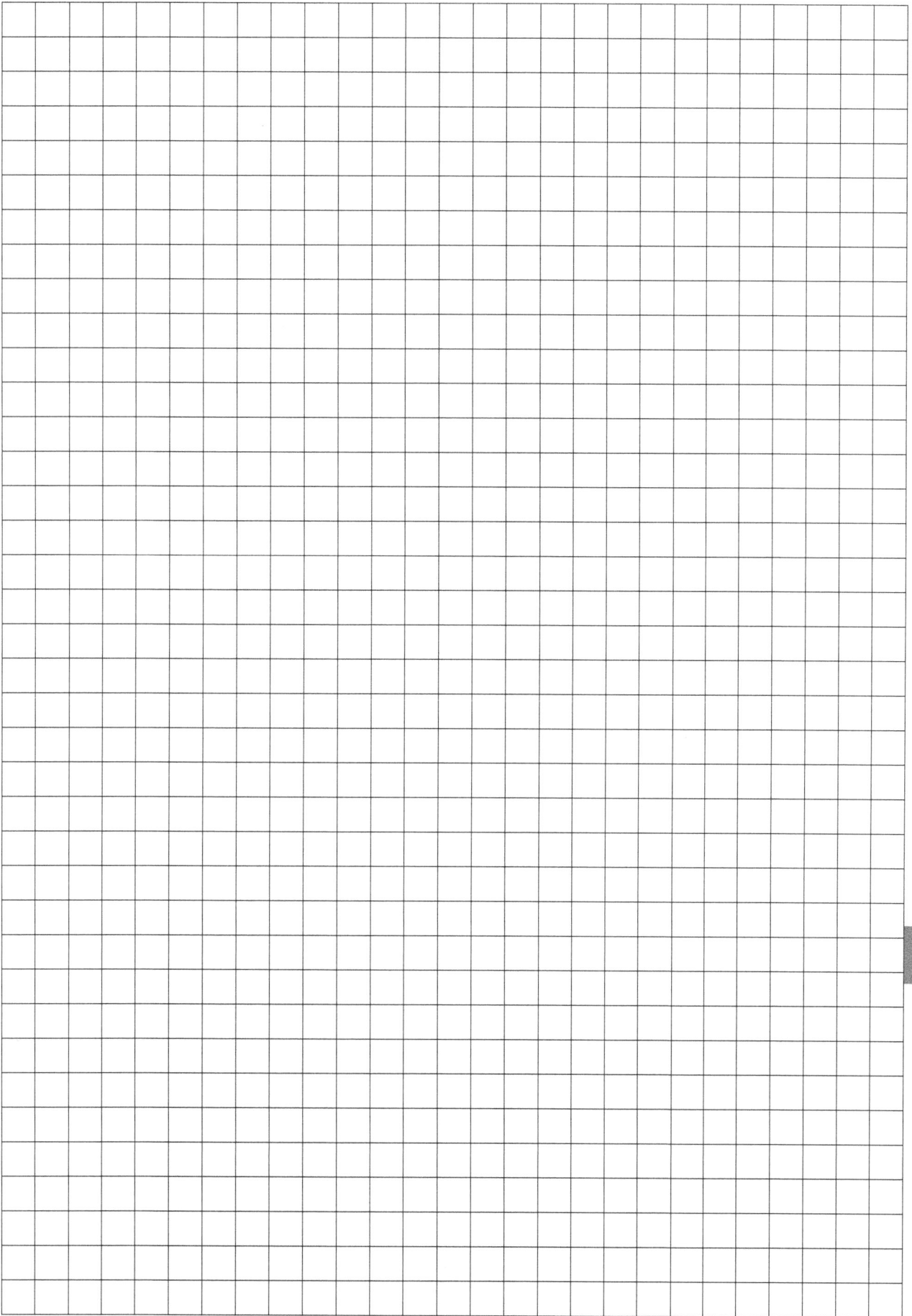

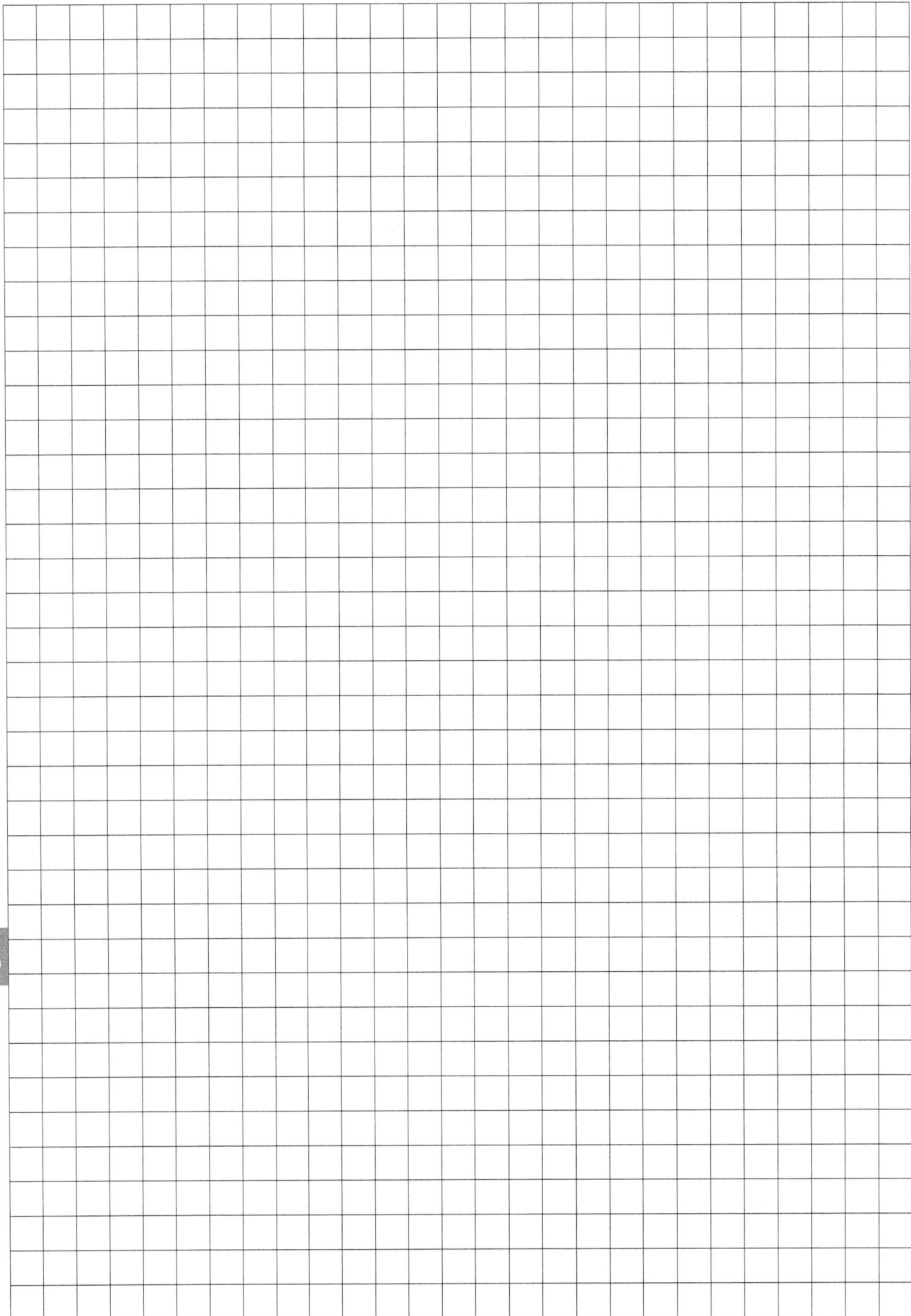

12 | Coasts with Google Earth

Purpose

In this lab, you will use Google Earth to study coasts of Oregon and other parts of the US.

Introduction

Coasts are the interface between land and sea. They make up a significant part of the Earth and exist on every massive continent and tiny island (and everything in between). Nearly 40% of the world's population lives within 60 miles of the coast and, because of their natural beauty and opportunities, people flock to visit them. Coastal geology, a subdiscipline of geology, delves into the intricate processes and distinctive features that characterize these areas. In this lab, we will closely examine various coastal locations, examining their specific features and important processes.

Part 1: Setup

Follow the directions below before answering the questions. Follow the instructions of your instructor to turn in any screenshots or digital images.

(Note: the first four steps below may be redundant from Lab 2 as your settings have probably not changed, but it is good to double check. Don't overlook the last two steps!)

1. Open Google Earth Pro (or download if using a new device; see Lab 2 for instructions).
2. Go to the 'View' option at the top of the screen. Make sure all these options are checked (figure 12.1):
    a. Toolbar
    b. Sidebar
    c. Status bar
    d. Scale legend

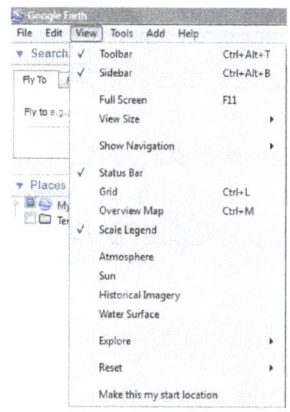

Figure 12.1. Google Earth Pro Required View Options

3. Declutter the map: Keep Borders and Labels, Places, and Terrain boxes checked. Uncheck all other options (figure 12.2).

Figure 12.2. Google Earth Pro Layers Options

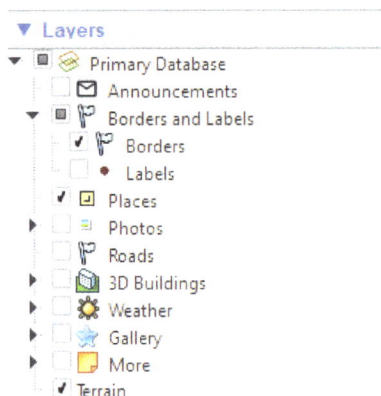

4. Make sure that latitude and longitude are displayed in degrees, minutes, and seconds and that units of measurement are in feet and miles and that the measurement units are English. Go to Tools→Options.
5. Download the Coasts kmz file (found in Canvas module) to your device.
6. Open the kmz file in Google Earth by clicking File>Open and select the kmz file. This will put placemarks (pins) on your map.

Part 2: East and Gulf Coasts of the US

Outer Banks, North Carolina

1. Navigate to this location using the search bar. View the entire stretch of islands across the state (roughly north to south).

1. The Outer Banks (as a whole) are what type of coastal feature?

2. Using the Path tool, make a path ~10 miles[1] long that goes across (perpendicular to) a section of the Outer Banks (anywhere between Kitty Hawk and Ocracoke). It should extend about equally on either side of the island.
3. We want to see the full form of the island you are profiling. To do this, base your elevation profile off of the seafloor, instead of sea level. Right-click on the path again and select Properties. Then click the "Altitude" tab and change from "Clamped to ground" to "Clamped to seafloor." Notice how your profile changed.

---

1   Remember: selecting the "Measurements" tab while drawing your path will show you the length of your path.

2. Take a screenshot that shows: an oblique view of the island, the profile line on the map, and elevation profile. Upload it, **following the directions given by your instructor**.

3. What is the maximum elevation (in feet) of the island in your profile?

4. Find an inlet (opening to the ocean), e.g., Hatteras, Oregon, Ocracoke. Observe the submerged sediment patterns that are on either side of the inlet. Describe their appearance and give an explanation for them.

5. Using the Historical Imagery tool, look at historical images (at least three) of the same inlet over the span of at least ~25 years. How has the appearance of these features changed over the decades? Be specific and reference specific times in your description.

4. Using the search bar, navigate to the Cape Hatteras Lighthouse. Add a Placemark and name it.

6. Using the Historical Imagery tool, go back to 2/1998. Here you will find the lighthouse in its original location (built in 1870) – look east and a bit north (the shadow will help you find it). Using the ruler tool, measure how far the lighthouse was from the shoreline (look for waves) in feet.

12

7. Using the ruler tool, measure the straight-line distance (in feet) from its previous location to its current location.

8. Explain why it was moved. (Recall from discussion or look it up.)

9. Zoom (way) out and find the Mid-Atlantic Ridge (it is not labeled but should be obvious based on our plate tectonics discussion). Using the ruler tool, measure the distance (in miles) from Cape Hatteras to the closest point on the MAR straight to the east.

Cape May Beach, New Jersey
1. Navigate to this location using the search bar (important: don't forget the word "beach" in your search). Zoom in and observe the shape of the beach.

10. What type of shoreline engineering do you observe? How can you tell? (Explain)

11. Based on your observations (and your answer above), what is the dominant direction of longshore current? Zooming in close to one or several of the structures will help.

12

Florida
1. Navigate to this location using the search bar and zoom out so that the entire state is visible.
2. Start by looking at the continental margin on the Atlantic side. Make a 300-mile path from **west to east** that starts in Melbourne. Make an elevation profile and clamp it to the sea floor (as you did in North Carolina above).

12. Take a screenshot that shows your profile line on the map and the elevation profile. Upload it, **following the directions given by your instructor**.

13. About how far offshore (in miles) is the shelf break from Melbourne?

3. Staying on the Atlantic side, find a pair of jetties surrounding an inlet. *(Hint: they are most abundant in the southern part of the state.)* Look at the shape/width of the beach on either side of the inlet and determine which is the "upshore" side and which is the "downshore" side, based on long shore current. Using two placemarks, create two pins to label the "upshore" and "downshore" sides.

4. Using the path tool, make a **red** arrow showing the direction of longshore transport.

14. Take a screenshot that shows the two labeled pins and the red arrow. Upload it, **following the directions given by your instructor**.

5. During the last glacial maximum (around 18,000 years ago), sea level was about 400 feet (125 m) lower than it is today. Using the Path tool and moving your cursor around the continental shelf surrounding Florida to observe water depth, draw a path around the peninsula (both sides) to estimate where this ancient shoreline was. Use click points every 50 miles or so. Color your path **yellow** and give it a weight of 2.0.

15. Take a screenshot that shows the path indicating the ancient shoreline. Upload it, **following the directions given by your instructor**.

16. Using the ruler tool, measure the current width (in miles) of the Florida peninsula from west to east at its widest point. *(Hint: This is roughly from Clearwater on the Gulf side to Melbourne on Atlantic side. Turning on "Places" in the Layers menu can help you find these locations.)*

12

17. Using the ruler tool, measure the ancient (18 Ka) width (in miles) of the Florida peninsula from west to east approximately along the same line.

18. How much narrower is the width of the Florida peninsula today than it was 18,000 years ago?

Part 3: West Coast of the US

Oregon

1. Navigate to Oregon (enter in search bar) and make your way to the coast, but zoom out so you can see beyond the continental shelf.
2. Using the path tool, make a 100-mile-long **west-to-east** path that starts 100 miles offshore and ends somewhere between Newport and Florence, Oregon (right at the shore), adjusting starting point as needed if needed. Make an elevation profile and clamp it to the sea floor.

19. Take a screenshot that shows the profile line on the map and the elevation profile. Upload it, **following the directions given by your instructor.**

20. Based on your profile, about how far offshore (in miles) is the shelf break from the Oregon Coast?

3. For questions 21–24 below, find the examples of the following features or phenomena anywhere on the coast of Oregon. Use the placemark tool and create a new pin. Give the name of each. If it doesn't have an official name, just title your placemark as the feature you are identifying. Turning on the "Places" layer may be very helpful in this section.

21. A headland (or "head" as they are often called here in Oregon). Take a screenshot with your labeled pin. Upload it, **following the directions given by your instructor.**

22. A small, well-defined bay protected by two headlands. Take a screenshot with your labeled pin. Upload it, **following the directions given by your instructor.**

23. A sea stack. Take a screenshot with your labeled pin. Upload it, **following the directions given by your instructor.**

24. Look at the waves now! Find a reasonably well defined rip current. Take a screenshot with your labeled pin. Upload it, **following the directions given by your instructor**.

4. Now we will look at Oregon's coast 18,000 years ago. Using the same technique as the Florida exercise, draw a path that indicates the location of Oregon's coastline when sea level was 400 feet (125 m) lower than it is today.

25. Take a screenshot that shows the path indicating the ancient shoreline. Upload it, **following the directions given by your instructor**.

5. Zoom out so that you can see the whole Northwest coast from an altitude ("eye alt") of about 1000 miles. Find the Juan de Fuca plate[2] based on the features on the seafloor and carefully highlight it using the Polygon tool. Change the color to **red** and adjust the area opacity to about 50%.

26. Take a screenshot that shows your Juan de Fuca polygon. Upload it, **following the directions given by your instructor**.

6. Unselect the polygon you just made from the Places menu so that it disappears from the map.
7. Find the spreading center[3] (i.e., the divergent plate boundary) between the Pacific Plate and Juan de Fuca plate straight west of Gold Beach, Oregon. Make a path from **west to east** that extends across the spreading center as perpendicularly as possible (about 125 miles long and centered over the axis of the spreading center). Make an elevation profile and clamp it to the seafloor.

27. Take a screenshot that shows the profile line on the map and the elevation profile. Upload it, **following the directions given by your instructor**.

28. From your elevation profile in the previous question, note that the spreading center (where the divergence magic is happening!) specifically is the V-shaped area that looks symmetrical on either side of it. Using the ruler tool, measure the distance (in miles) from Gold Beach to the middle of the spreading center.

12

---

2 Hint: go back to Plate Tectonics chapter or discussion if you need some help here. Include Gorda and Explorer portions of plate.

3 Review images of spreading centers/MORs to remind yourself of what this looks like. It should be the western boundary of the southern part of JdF plate polygon.

Part 4: East and West Coast Comparison

29. Compare and contrast the east coast locations with those of Oregon/west coast. What are some major differences? Discuss features, sizes and shapes of the continental shelves (and how well defined they are), distance to plate boundaries, etc. This should be a paragraph – at least 4 full sentences.

30. Revisit your images of the ancient (18 Ka) shorelines of both Oregon and Florida. Compare the two and explain why they are significantly different from each other.

31. Is the east coast a submergent or emergent coast?

32. Is the west coast a submergent or emergent coast?

13 Deserts

Purpose
To provide an overview of common desert features and the environment where they form.

Materials
- ❑ Three Forks Quadrangle
- ❑ Chromebook

Part 1: Name that Dune

For each figure in this section:
1. Name the type of dune.
2. Label the slip face of the dune.
3. Draw arrows indicating the dominant wind direction or directions.

Figure 13.1.

1. Name of the dune _____

Figure 13.2.

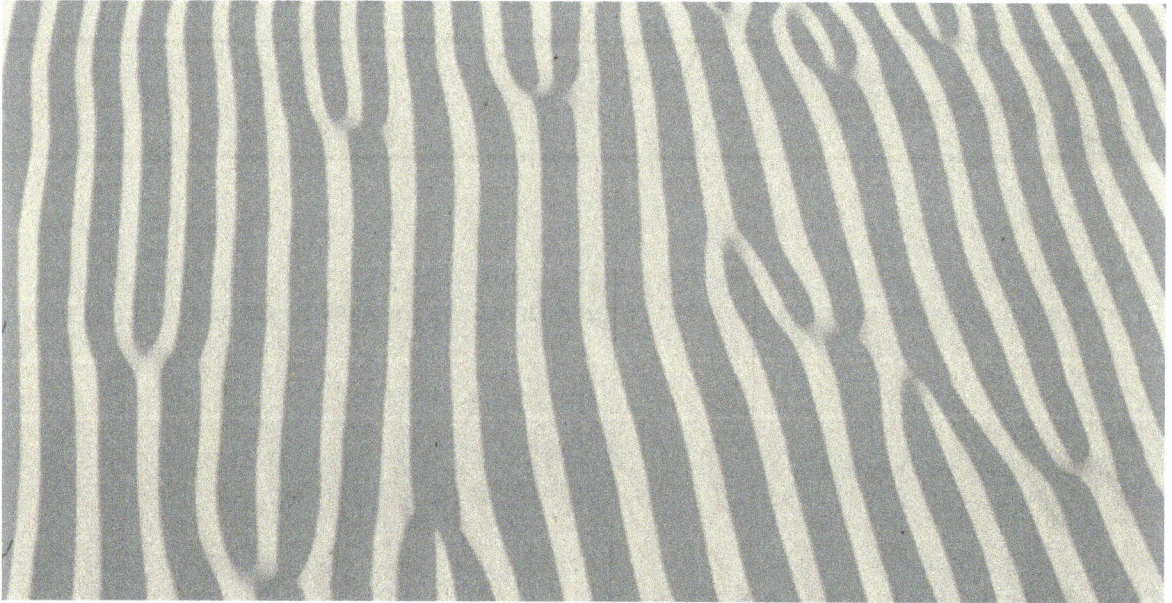

2. Name of the dune _____

Figure 13.3.

3. Name of the dune _____

13

Figure 13.4.

4.   Name of the dune _____

Figure 13.5.

5.   Name of the dune _____

Part 2: Deserts of North America

Use figure 13.6 to answer the questions in this part.

Figure 13.6. Steppes and Deserts in the Continental United States

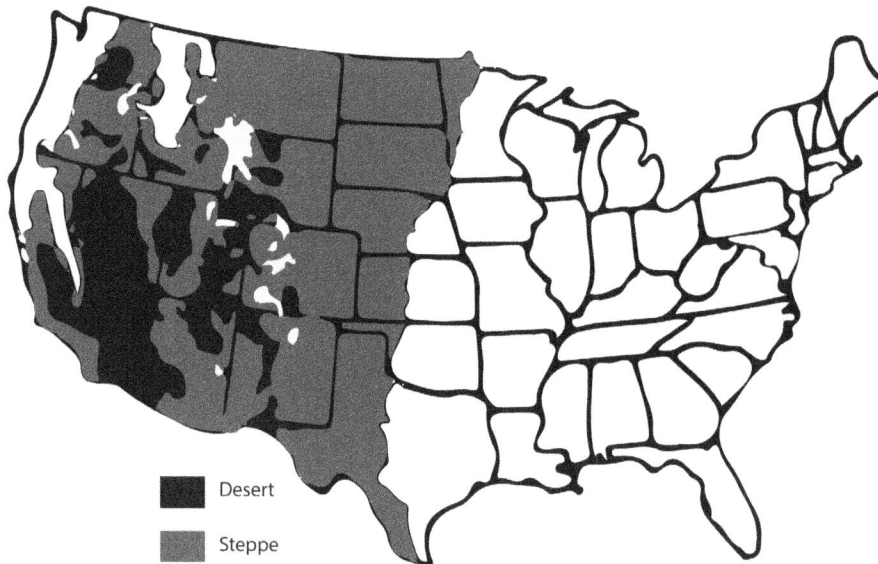

6. In what area or state of the United States does the greatest area of true desert exist? What is the most likely cause of deserts at this location (i.e. atmospheric circulation, the locations of mountain, etc.)

7. Why would there be such a large extent of desert and steppe in the western US but not the eastern US? Be specific in your answer. Obviously one area gets more precipitation, but why?

8. What feature is directly responsible for the dry areas of eastern Oregon and Washington?

Part 3: Maps of Desert Areas

Three Forks Quadrangle

Look at the Three Forks Quadrangle and answer the following questions to the best of your ability. Use the map and what you know about surface processes.

9. Describe the overall topography. Explain both the high and low elevations of the map. Write at least a paragraph.

10. What are the light red dotted lines on the map (check the key)?

11. Based on the fact that this is a canyon, do you think the canyon walls are made of sediment or hard rock? Explain your answer.

12. Look at the Middle Fork of the Owyhee River. What does the dotted blue line mean? What specifically are these types of streams called?

13. Look at the lakes on this map. How do you think these lakes are formed and why do you think that? What would these lakes be used for?

14. Look just north of where Porcupine Canyon enters the mainstem Owyhee. On the right side of the river is a landslide (that backwards "C" shape of the canyon walls). Based on what you know about meandering rivers, how do you think that landslide formed?

15. There are at least two terraces that can be found in the canyon in the northern portion of the map. What is the general elevation of these terraces (look near the word "Deary")?

13

16. Compare the satellite image and the topographic map of Jackson Hole (figure 13.7). Notice that streams feed into the main river from both the east and west. How are the contour lines on the east side of the map different from a typical stream pattern (seen on the west side)?

Figure 13.7. Topographic map and satellite images of Jackson Hole, Oregon.

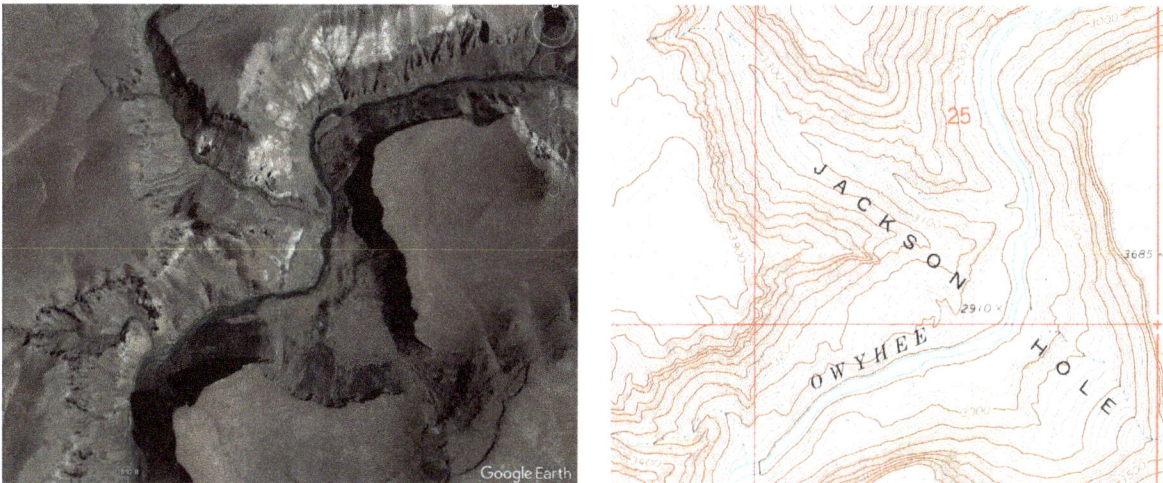

17. Based on the map and satellite image in figure 13.7, what common desert features are forming on the east side of Jackson Hole? Explain your reasoning.

18. Measure the slope (gradient) of the eastern slope (note: the map scale is 1:24000). Show your work.

Christmas Valley Sand Dunes with Google Earth
Using Google Earth, look at the Christmas Valley Sand Dunes in south-central Oregon.

19. You should notice several lakes in the area (such as Summer Lake to the south). Provide at least two pieces of evidence that these lakes evaporate a significant amount during the dry months of the year.

20. These lakes were part of a larger system of lakes that formed during glacial periods. What is the name for this type of lake and how do they form?

13

21. What type of dunes are the main dunes in the Christmas Valley Sand Dunes area? Explain your reasoning.

22. The sediment in the Christmas Valley Dunes is volcanic. What do you think is the source for the sediment that makes up the dunes?

13

14 Deserts with Google Earth

Purpose

In this lab, you will use Google Earth to study deserts and drylands in Oregon and other parts of the world.

Part 1: Setup

Follow the directions below before answering the questions. Follow the instructions of your instructor to turn in any screenshots or digital images.

(Note: the first four steps below may be redundant from Lab 2 as your settings have probably not changed, but it is good to double check. Don't overlook the last two steps!)

1. Open Google Earth Pro (or download if using a new device; see Lab 2 for instructions).
2. Go to the 'View' option at the top of the screen. Make sure all these options are checked (figure 14.1):
   a. Toolbar
   b. Sidebar
   c. Status bar
   d. Scale legend

Figure 14.1. Google Earth Pro Required View Options

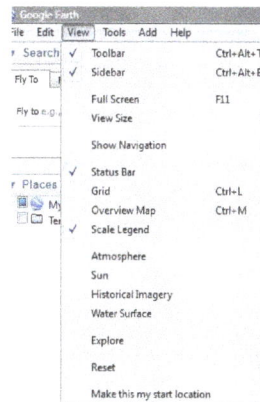

3. Declutter the map: Keep Borders and Labels, Places, and Terrain boxes checked. Uncheck all other options (figure 14.2).

Figure 14.2. Google Earth Pro Layers Options

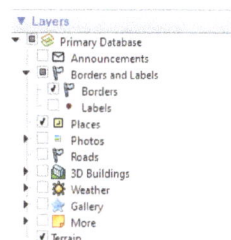

4. Make sure that latitude and longitude are displayed in degrees, minutes, and seconds and that units of measurement are in feet and miles and that the measurement units are English. Go to Tools→Options.
5. Download the Deserts kmz file (found in Canvas module) to your device.
6. Open the kmz file in Google Earth by clicking File>Open and select the kmz file. This will put placemarks (pins) on your map.

Part 2: Deserts of the World

Sahara Desert, North Africa

1. Navigate to the following coordinates using the search bar: 21°23'4"N 11°59'00"E
2. Zoom out so that the entire Sahara Desert is visible (as roughly defined by the tan color/ absence of vegetation).

1. Name three countries that primarily lie within the Sahara Desert.

2. Using the Ruler tool, determine the length of the Sahara Desert from west to east in miles (roughly in the middle).

3. Using the Ruler tool, determine the length of the Sahara Desert from north to south in miles (from Tripoli, Libya to the Niger/Nigeria border).

4. What is the area of the Sahara Desert (according to *your* measurements)?

5. Make similar measurements of the contiguous US (i.e., not including Alaska and Hawaii) and compare them to the dimensions of the Sahara Desert.

14

3. Using the Polygon tool, highlight (trace) the Sahara Desert. Do this by clicking on the perimeter until it covers the tan part.
4. In the "Style, Color" tab, select **yellow** for the area color and change its opacity to 50%.

6. From the "Measurements" tab, determine the area of the Sahara Desert (in square miles) based on your polygon.

5. Give your polygon a name and click OK.

7. Take a screenshot that shows your polygon. Upload it, **following the directions given by your instructor**.

8. How did your calculated value for area (question 4) compare to your measured value for area (question 6)?

9. What is the latitude range of the Sahara Desert? Give the approximate latitudes[1] of the southern and northern margins of the Sahara Desert.

10. Based on your previous answer, what type of desert is the Sahara Desert?

11. What is the reason for this area being a desert?

14

---

1   Reminder: hover your mouse over a specific point and latitude, longitude, elevation, etc. will appear in the bottom of your screen.

Gobi Desert, East Asia

1. Navigate to the following coordinates using the search bar: 43°02'18"N 106°12'25"E. Then zoom out a bit and look at the topography.

12. Although there is no obvious or substantial water to be seen, describe evidence or features that you see that appear to be caused or formed by running water.

2. Check the "Borders and Labels" box in the Layers menu.

3. Zoom out so that the entire Gobi Desert is visible (as roughly defined by the tan color, though not as obvious as the previous example). You may also be able to hover your mouse over the term "Gobi Desert" to see the extent.

13. Name the two countries that make up the most significant portion (nearly all) of the Gobi Desert.

14. What is the approximate latitude range of the Gobi Desert?

15. Based on your previous answer, what type of desert is the Gobi Desert?

16. What is the reason for this area being a desert?

14

Part 3: Deserts of the United States

Death Valley National Park, California

1. Navigate to this location using the search bar. Zoom in and look around, observing the topography and desert features.
2. Find the single (larger) alluvial fan. There are quite a few with obvious shape and appearance. Using the Path tool, highlight[2] (trace) around it. Change the color to **red** and its width to 2.0.

17. Take a screenshot that shows your highlighted alluvial fan. Upload it, **following the directions given by your instructor**.

3. Find a bajada (there are many in the general area). Using the Path tool, trace it. Change the color to **dark blue** and its width to 2.0.

18. Take a screenshot that shows your highlighted bajada. Upload it, **following the directions given by your instructor**.

4. Find a salt pan. Using the Path tool, trace it. Change the color to **black** and its width to 2.0.

19. Take a screenshot that shows your highlighted salt pan. Upload it, **following the directions given by your instructor**.

20. In what American desert is Death Valley located? See figure 14.3.

Figure 14.3. Major North American Deserts

---

2 The Path tool is not just for creating elevation profiles. It can also be used to highlight or draw, in which case, elevation profiles are not necessary (unless otherwise indicated).

14

21. In what US geologic province is Death Valley located? See Figure 14.4.

   Figure 14.4. US Geolgical Provinces

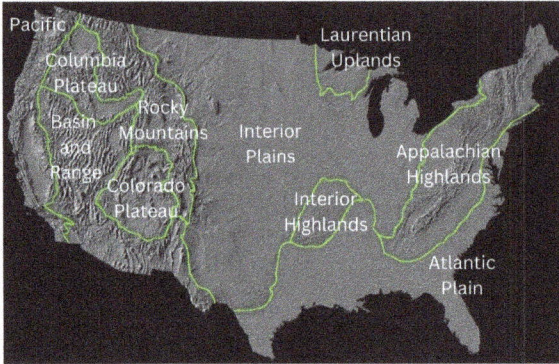

22. Zoom out further so that both Death Valley and Fresno (to the west) are visible. In what type of desert is Death Valley located?

23. What is the reason for this area being a desert?

Great Sand Dunes National Park & Preserve, Colorado

1. Navigate to this location using the search bar. Zoom in to look at the roughly crescent-shaped dune field.

24. Using the Ruler tool, measure the width of the dune field (roughly north to south at its widest).

2. Navigate to the following coordinates using the search bar: 37°48'14" N 105°34'35" W. Zoom in until about four adjacent dunes are in close view.

3. Adjust the perspective to an oblique view so that you can get a good sense of the shape of the dunes.

4. Draw a path across three or four adjacent dunes and make an elevation profile. Be sure that three or four crests are visible.

25. Take a screenshot that shows the profile line on the map and the elevation profile. Upload it, **following the directions given by your instructor**.

26. Based on the dune appearances and your elevation profile, what type of dunes are shown in this area?

27. Based on the dune appearances and your elevation profile, from what direction does the wind blow in this area most consistently?

28. This dune field is considered a draa. Define what is meant by a draa (in your own words) and explain why this area is considered a draa.

Part 3: Deserts and Dry Lands of Oregon

Although Oregon is probably best known for its substantial rainfall and temperate rainforests, dry lands actually make up the majority of the state's area. Here, we will look at some of the desert features of Oregon's dry lands.

The Greater Pacific Northwest

1. Navigate to Oregon[3] and zoom out so that the greater Pacific Northwest area is shown.
2. Observe the vegetation differences throughout this area and visually identify where the dry lands mostly are mostly located.
3. Using the Path tool, trace rain shadow line of the US Pacific Northwest (not just Oregon). Change the color to **white** and adjust the width to 2.0.

29. Take a screenshot that shows your highlighted path. Upload it, **following the directions given by your instructor**.

14

---

3   Turning on Borders in the Layers menu (if not already on) will help you find and center the Pacific Northwest.

4. Make an elevation profile that extends from Sweet Home to Redmond, Oregon, (**west to east**) and observe how the topography and elevation change along the profile line.

30. Take a screenshot that shows the profile line on the map and the elevation profile. Upload it, **following the directions given by your instructor**.

31. What is the average annual rainfall amounts in these two locations (in inches)?

32. Describe the rain shadow effect (in your own words) and use it to explain why there is such a great difference in annual rainfall between these two specific locations.

Christmas Valley Sand Dunes

1. Navigate to this location[4] using the search bar. Zoom out so you can see its entire extent, then zoom in more closely so that you can observe the shape of the sand dunes.

33. These sand dunes have a fairly unusual composition. Look up and describe the composition of the sand making up these dunes.

2. Make an elevation profile that extends across 3 to 4 adjacent dunes and observe the shapes.
3. Look closely at the slopes of the dunes in your profile and determine where the stoss versus slip faces are.

34. Take a screenshot that shows the profile line on the map and the elevation profile. Upload it, **following the directions given by your instructor**.

4. Look at the slopes shown in your profiles and identify the stoss and slip face sides.

---

4   Not the town of Christmas Valley, Oregon.

35. Which side of the dunes (on average) is the stoss side?

36. Which side of the dunes (on average) is the slip face?

37. What is the average height of the dunes in your profile (in feet)?

38. What is the prevailing wind direction in this area?

39. Based on your observations and responses to the previous questions, what type of dunes are these?

The Alvord Desert

1. Navigate to this location using the search bar. Zoom out so you can see its entire extent.

40. This is the driest place in Oregon! What is its annual rainfall amount (in inches)?

2. This is also one of the flattest expanses of land in Oregon! Draw a path from **west to east** across the widest part of the Alvord Desert. Make sure to start and end your profile just beyond the margins of the Alvord Desert (getting just slightly into the darker colored areas on each side). Make an elevation profile for this path and observe the shape of the Alvord Desert compared to the surrounding land.

14

41. The Alvord Desert is quite small and, as a whole, is actually two related desert features. Based on its appearance, size, and profile, what types of desert features compose the Alvord Desert?

42. Take a screenshot that shows the profile line on the map and the elevation profile. Upload it, **following the directions given by your instructor**.

43. In which Oregon geological province is the Alvord Desert located? See figure 14.5.

Figure 14.5. Oregon Geological Provinces

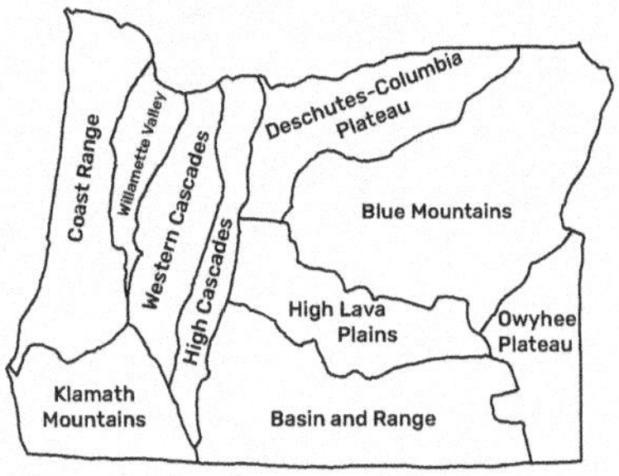

Summer Lake

1. Navigate to the following coordinates using the search bar: 42°50'00"N 120°45'13"W. This is a similar feature to the Alvord Desert except that there is (some) water in Summer Lake.

44. Observe the area closely in overhead, oblique, and ground view. In what ways (indicate at least 2) does this look similar to the Alvord desert?

45. These are both former pluvial lakes. What are pluvial lakes and how do they form?

2. Find three other features (besides Summer Lake and the Alvord Desert) in the area that are also former pluvial lakes and trace them using the Path tool. Change the color to **red** and width to 2.0.

46. Take a screenshot that shows the three highlighted features. Upload it, **following the directions given by your instructor**.

47. These (and many others in the area) were once a part of what Pleistocene Lake?

14

15 Landslides and Mass Wasting

Purpose

To identify types of mass wasting events and explore the factors that influence mass wasting events.

Materials

- ❑ Flat board
- ❑ Protractor
- ❑ Friction block
- ❑ Various weights
  (that fit in the friction block)

- ❑ Sand
- ❑ Rice
- ❑ Paper plate or pie tin
- ❑ Gravel

Part 1: Name That Slide

1. Earlier, you learned about the different types of landslides. What type of mass wasting event is described or shown in figure 15.1? Choose from the following:

   Solifluction | Soil Creep | Slump | Rockfall | Rockslide | Debris Flow

   Figure 15.1.

   Name of wasting event: _____

2. Which type of mass wasting event occurs when blocks of layered bedrock have broken loose and moved downslope, usually along a weakened fracture or boundary between layers? _____

3. Identify the wasting event in figure 15.2

Figure 15.2.

Name of wasting event: _____

4. Identify wasting event in figure 15.3.

Figure 15.3

Name of wasting event: _____

5. Identify the wasting event in figure 15.4.

   Figure 15.4.

   Name of wasting event: _____

6. Which type of mass wasting event occurs when saturated soil moves over permafrost, leading to a lumpy, folded landscape? _____

Part 2: Things Are Moving

For this part of the lab you will be examining the effect of mass and gravity on the critical angle for a slope (the angle when sliding starts).

Procedure

1. Use the board as an inclined plane (make a ramp). Align the bottom center of your protractor with the corner of the ramp so that you can see the numbers on the protractor.
2. Place the friction block on the higher end of the board (opposite the protractor).
3. Slowly and steadily lift the board and friction block. Stop lifting when the block starts to move. Record the angle of the inclined plane shown on the protractor.
4. Repeat 10 times and fill out table 15.1: Weight A Trial Chart.
5. Find the average angle and place it in the blank below the table.
6. Place additional weight in the friction block. Try to set it up so the weight will not shift while the inclined plane is being lifted. Repeat steps 1–5 with the heavier friction block and table 15.2: Weight B Trial Chart.

Table 15.1: Weight A Trial Chart

Weight A: _____ g										
Weight A Trial	1	2	3	4	5	6	7	8	9	10
Angle										

The average critical angle for the empty friction block is: _____

Table 15.2: Weight B Trial Chart

Weight B: _____ g										
Weight A Trial	1	2	3	4	5	6	7	8	9	10
Angle										

The average critical angle for the empty friction block is: _____

7. Were the critical angles different or not different? Why or why not?

8. What do you think would happen to the critical angles if the board were covered with water? Why?

Part 3: The Angle of Repose

In this part you will explore the angle of repose for various sediments: dry sand, damp sand, saturated sand, gravel, and rice.

Procedure

1. Write a hypothesis. Which material will make the steepest slope and why? Which will make the shallowest and why?

    Hypothesis

2. On a paper plate or pie tin, make a small pile of the first sediment. Keep piling the material until it is continuously falling downslope and you can't build it any steeper. *Note: make your piles on a flat surface, not the ramps from Part 2.*
3. Measure the angle created by the table and the pile's slope. Record your answer in table 15.3.
4. Repeat with each material.

Table 15.3: Angle of Repose

Material	Dry Sand	Damp Sand	Saturated Sand*	Gravel	Rice
Angle					

* "Saturated" sand means that it is soupy or liquid.

9. Was your hypothesis about the steepest material correct? Which material made a steeper angle? Explain why.

10. Was your hypothesis about the shallowest material correct? Which material made a smallest angle? Explain why.

Part 4: Possibility of Mass Wasting

The map in figure 15.5 was produced by the USGS and shows the areas most susceptible to land-slides. According to the map, some places have no landslide susceptibility (tan) while other places have a very high landslide susceptibility (red). Answer the following questions based on the information provided by figure 15.5.

Figure 15.5. USGS Map of Landslide Risk

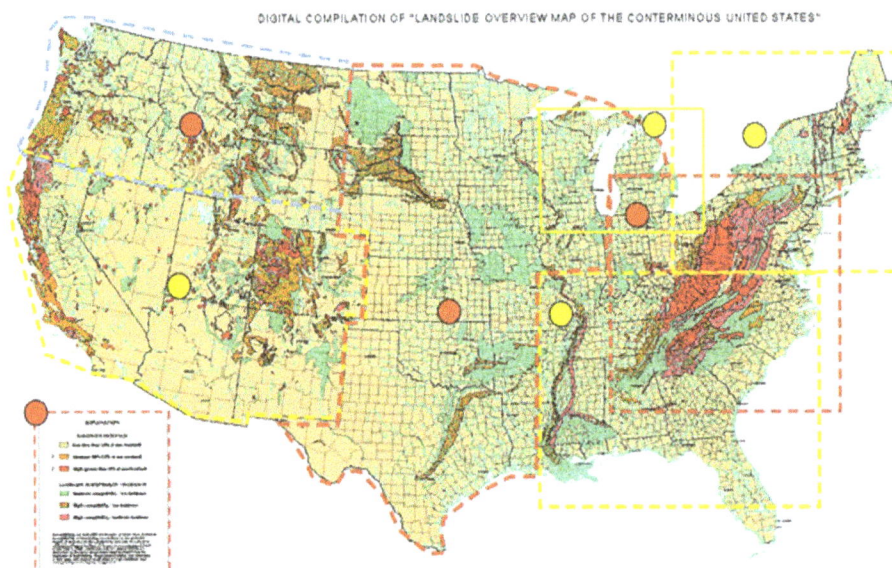

11. Hypothesize which types of terrain and climate are needed for a greater possibility of landslides. Write your hypothesis here.

12. Why is the western portion of Oregon more susceptible than the east side?

13. Why is there a high susceptibility of mass wasting in Colorado? What sort of rocks are present there and why would these rocks induce landsliding?

14. There is a huge area in the east that has a high landslide susceptibility. Why would this area show high susceptibility and what is the geology in that area?

Part 5: Oso Landslide

Use the images in figure 15.6 to answer questions about the Oso Landslide.

Figure 15.6. Aerial Images of Oso Landslide

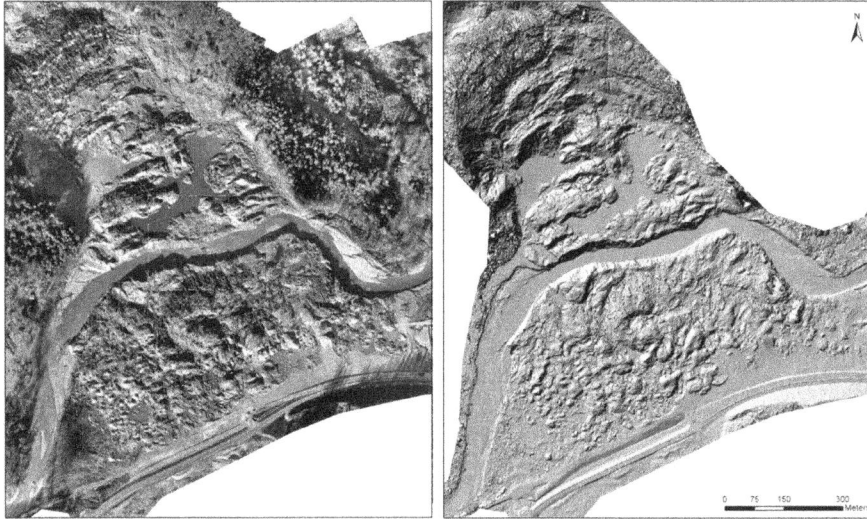

15. The photos in figure 15.6 are an aerial photo (left) and LiDAR (right) of the Oso Landslide. Using both, what would you name this type of landslide? *Hint: this is a landslide complex, so you can see a different type of slide near the scarp versus the toe. Try to name both.*

16. Now you will calculate the speed of the landslide. First, measure the distance of the landslide from the scarp to the toe using the scale provided on the LiDAR picture. The actual movement of the landslide took 42 seconds. Calculate the speed in meters per second (m/s) and miles per hour (mi/hr). *Note: there are approximately 3.28 ft in 1m.*

17. The aerial photo (on the left) was taken after the landslide occurred. Notice the river runs directly through the body of the landslide now. What would have happened to the river as the landslide initially crossed the valley? How could this have affected people both upstream and downstream of the event?

16 Oregon Surface Geology in Google Earth

Purpose

To use Google Earth (or Google Maps with the satellite view) to identify various features in Oregon from topics covered throughout the quarter. This is an opportunity to review for the final exam.

Instructions

Using Google Earth, explore Oregon and try to identify the various features you have learned about this quarter. You will take screenshots of these. Follow the instructions of your instructor to turn in any screenshots or digital images. Use the following rules to select your features:

1. The feature must be from Oregon.
2. You need **at least 3 of each** of the following:
   a. Mass Wasting Features (scarps, debris flows, rockslides, etc.)
   b. Stream Features (oxbow lakes, point bars, deltas, etc.)
   c. Glacial Features (horns, troughs, tarns, etc.)
   d. Desert Features (yardangs, alluvial fans, playas, dunes, etc.)
   e. Coastal Features (sea stacks, spits, marine terraces, etc.) Note: for this assignment, dunes will only apply as a desert feature.
3. The location cannot be an example we used in lecture or lab already this quarter.
4. Do not use the same feature more than once (don't submit three images of "glacial U-shape valley").
5. All of Oregon's geomorphic provinces should be represented with the following adjustments:
   a. The Willamette Valley, Coast Range, and Klamath Mountains can be lumped together as one province.
   b. The Western and High Cascades can be lumped together as one province.
   c. The Columbia River Plateau will remain its own province.
   d. The High Lava Plains, Owyhee Highlands, Basin, and Range will all be one province.
   e. The Blue Mountains will remain its own province.
6. The feature must be a landform created by natural deposition or erosion, and not a man made construct. Please include specific landforms created by the various processes discussed in class. **Do not** simply include a picture of a stream and say "meandering stream." It must be a specific example of a feature made by the meandering stream, such

as a point bar, oxbow lake, or river terrace. Don't simply post "glacier" or "landslide" or "Oregon Desert." Look for specific examples of landforms created by these processes.

For Your Write Up

Include the following in the provided formatted file, neatly organized and printed or emailed to your instructor. Please try to make the pictures as small as possible while still showing the needed feature.

1. Copy or screenshot each feature. Do not use street view or photos. Use aerial/satellite pictures from Google only. Tilt the view if you wish but try not to distort it too much.
2. As typed text in the document, label the feature with the following:
   a. What the feature is, or at least what you think the feature is
   b. What process created the feature (coast, stream, glacier, desert, mass wasting)
   c. Is the feature made by erosion or deposition (if both, which one dominates)
   d. The Latitude/longitude of the feature: this can be found in Google Maps or Google Earth in several ways. Use internet searches or ask your instructor if you are not sure how to find it. Please use decimal degrees or degrees-minutes-seconds.
   e. The Oregon Province where the feature is located

17 Field Trip: Missoula Flood Features

Stop 1: Erratic Rock Natural Site

1. What type of rock is this? Is it metamorphic, igneous, or sedimentary? Describe its texture, color, and features.

2. What are the dimensions of this rock in centimeters? Record its dimensions. It is currently 90 tons, but it was originally 160 tons prior to people chipping away at it. Whoa!

Figure 17.1. Velocity of Water in Moving Boulders

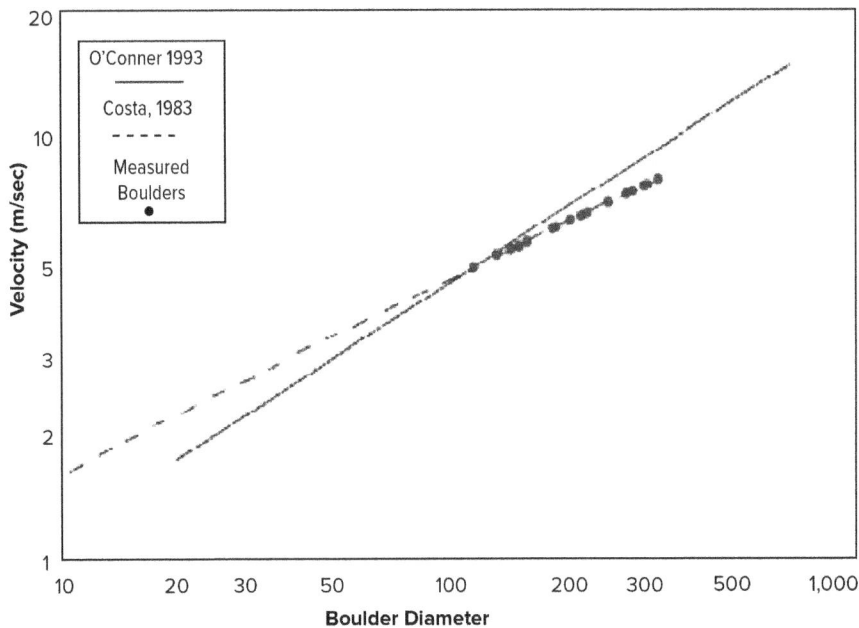

3. Based on your measurements and the graph in figure 17.1, how fast would water have to move to be able to lift this boulder and move it? Just for comparison, the Willamette's flow is generally .3m/s. These erratics were deposited in another way. How?

4. Look out over the valley. Can you see any other features that would lead you to believe a large flood swept through here?

Stop 2: Tualatin Mastodon

5. What exactly is a mastodon? What is its most likely relative that is still alive today?

6. Notice these other erratics. What types of rocks do you think these represent? Where would we find similar rocks today?

Stop 3: Willamette Falls

7. What type of rock do you think makes up these falls? How were these falls carved out? What evidence can you see in the valley walls that supports your hypothesis?

Stop 4: Camassia Natural Site

8. What type of rock is this? What features of these boulders show evidence for water erosion? Sketch and label a boulder to help make your case.

9. This is a kolk (which is pretty amazing, considering we are so close to a city). How exactly do you think a feature like this could form?

Stop 5: Mount Tabor

10. Mt. Tabor is a cinder cone of the Boring Lava (Volcanic) Field. Take a look at this outcrop and sketch it below. How do you think these layers were created?

Figure 17.2. Height of Lake Allison from Mount Tabor

11. Now that we've walked to the overlook, check out the photo in figure 17.2. How high do you think those waters are based on our current elevation? Based on the map you've seen today, what about this area's topography would have caused the waters here to form a temporary lake that would have taken several days to drain?

18 Field Trip: Silver Falls State Park

Purpose
To observe and describe stream processes, especially waterfalls. Lots of waterfalls.

Methods
- ☐ Closed toed shoes and weather-appropriate clothing
- ☐ Pencil
- ☐ Water

Instructions
Begin at the overlook to South Falls.

Stop 1: Overlook to South Falls

1. Look into the gorge. What rock type do you think makes up the cliff?

2. Notice how the trail goes behind South Falls. Suggest a reason why this portion of the cliff has eroded into a cave/overhang.

3. Why do you suppose the dropoff is so much larger than the falls themselves?

Stop 2: Bridge in Front of South Falls

4. Sketch the falls. You will be graded on detail, not artistic ability. It should be obvious without a title that you have drawn South Falls. Be sure to include a scale.

Stop 3: Lower South Falls

Lower South Falls is about 1 miles further down the trail.

5. On the way to Lower South Falls you will pass evidence of mass wasting. What two types of mass wasting events do you observe? What evidence suggests why they occurred?

6. Draw a sketch of Lower South Falls. Don't forget to include a scale.

7. What is the shape of the valley? What does this suggest about the origin of the canyon?

8. What similarities do you see between the rock layers here at Lower South Falls and the layers at South Falls? Are they the same basalt flows? Why or why not?

Stop 4: Bend in the River Further Downstream

Find the series of boulders placed right along the trail where the river flows very close.

9.   Describe the flow of the river in this location. Is it a straight path? What portion of the channel seems to have the fastest flow?

18

10.   The large rocks are not natural, they have been moved there by the park. Why?

Stop 5: Toward Lower North Falls

As you head toward Lower North Falls, stay left when the trail splits. Along the way, find the small waterfall dropping into the canyon. This may be dry in the summer or fall, but the channels and dropoff are fairly noticeable (0.5–0.75 miles after the split).

11.   Are you walking upstream or downstream now? Is this the same stream we were just with at South and Lower South Falls?

12. Sketch the small waterfall. Be sure to catch the smaller channels' relationship with Silver Creek.

13. While smaller, this tributary has a well-defined channel. Why is its channel not as low as Silver Creek's? There is a term for this feature (a side valley situated above the main channel) which may help answer this question.

Stop 6: Bridge Over the Deep Pool Before Lower North Falls

14. Why is this section of the stream so much deeper than the rest (this is not the first deep pool you've come across)?

Stop 7: Lower North Falls

15. Sketch the falls. Last one, I swear.

16. How is this waterfall different from the others? Given your observations of the previous waterfalls and layers, suggest a reason for this waterfall's shorter height.

> Continue home. The remainder of the trail is for exploration and further observation. The Winter Falls Trail connects about a half mile up the canyon. This will take you back up the Rim Trail, which can be followed back to South Falls.

Acknowledgments

Text Acknowledgments

"Field Trip: Silver Falls State Park" by Michelle Harris, is copyrighted work and used by permission in this work only. It may not be used outside of this work or in any derivative work without further permission.

Image Acknowledgments

Figure 1.1. "Columbia River Terraces" by Shannon Othus-Gault is a product of work by Chemeketa Press.

Figure 1.2. "Steens U-Shape" by Shannon Othus-Gault is a product of work by Chemeketa Press.

Figure 1.3. "East Spring Earthflow" by Shannon Othus-Gault is a product of work by Chemeketa Press.

Figure 1.4. "Dunes" by Shannon Othus-Gault is a product of work by Chemeketa Press.

Figure 2.5 "USGSlabels, AI upscaling: Hike395" is in the public domain and is available via Wikimedia Commons.

Figure 2.6. Google Earth Pro. (July 30, 2016.) Snake River Canyon, 45°33'42.27"N, 116°27'09.89" W, Elevation 3994 ft, Eye alt 18878 ft.

Figure 2.7. Google Earth Pro. (July 21, 2024.) Chief Kiawanda Rock, 45°13'03.69" N, 123°58'12.83" W, Eye 4491 ft. Data SIO, NOAA, US Navy, NGA, GEBCO. Image © 2025 Airbus.

Figure 2.8. "Geomorphic Provinces of Oregon" is a product of work by Chemeketa Press.

Figure 4.1. "Bennett Pass Topographic Profile Graph" by Cierra Maher is a product of work by Chemeketa Press.

Figure 4.2. "Barlow Pass Topographic Profile Graph" by Cierra Maher is a product of work by Chemeketa Press.

Figure 5.1. "River Profile" by Cierra Maher is a product of work by Chemeketa Press.

Figure 5.2. "Hjulstrom Diagram" by Steven Earle is used under a Creative Commons Attribution 4.0 International (CC BY 4.0) licence. https://pressbooks.openeducationalberta.ca/practicalgeology/

Figure 5.3. "Homemade Half-pipe" by Shannon Othus-Gault is a product of work by Chemeketa Press.

Figure 6.3. Google Earth Pro. (September 11, 2015.) Yellowstone River, 44°43'27.83" N, 110°28'41.33" W, Elevation 7378 ft, Eye alt 16381 ft.

Figure 6.4. "Gradient Diagram" is a product of work by Chemeketa Press.

Figure 6.5. "Sinuosity Diagram" is a product of work by Chemeketa Press.

Figure 7.1. "Grid Overlay over Wah-Chang "Spill" area" by Google Maps is under copyright and used with permission (http://www.google.com).

Figure 7.2. "Satellite image of ATI Wah Chang Area" by Google Earth is under copyright and used with permission (http://www.google.com).

Figure 8.1. "Groundwater Contamination Map" by Chemeketa Press is a product of work by Chemeketa Press.

Figure 10.1. "Setup of Glacial Goo Activity" by Shannon Othus-Gault is a product of work by Chemeketa Press.

Figure 10.2 "Boulder Glacier" is in the public domain. Glacier National Park, Montana, Repeat photography 1910–2007. https://www.sciencebase.gov/catalog/item/5efb871182ced62aaaf058b0 Left photo summary: Boulder Glacier, circa 1910. Morton J. Elrod, photographer. Courtesy of Glacier National Park Archives. Right photo summary: Boulder Glacier from ridge near Chapman Peak, Glacier National Park, Montana. August 24, 2007, photo by D. Fagre and G. Pederson, USGS.

Figure 10.3. "Grinnell Glacier" is in the public domain. Glacier National Park, Montana. Repeat photography 1910–2016. Summary of left photo: Grinnell Peak. Glacier National Park, Montana. Photo taken August 9, 1910. M.J. Elrod, photographer. Courtesy of University of Montana, Mansfield Library Archives and Special Collections. Summary of right photo: Grinnell, Gem and Salamander Glaciers from Grinnell Ridge, Glacier National Park, Montana. Photo taken September 27, 2016 by Lisa McKeon, USGS. https://www.sciencebase.gov/catalog/item/5f2c6ecd82ceae4cb3c2d090

Figure 10.4. "Grinnell Glacier" is in the public domain. Glacier National Park, Montana. Repeat photography 1938 - 2019. Summary of left photo: Grinnell Glacier, Garden Wall in the center, Swiftcurrent Mtn. in the background, 1938. T. J. Hileman, photographer. Courtesy of Glacier National Park Archives. Summmary of right photo: Grinnell Glacier, Glacier National Park, Montana. Photo taken September 4, 2019 by Lisa A. McKeon, USGS. https://www.sciencebase.gov/catalog/item/5f457c7f82ce4c3d12251d64

Figure 13.1. "Barchan Dunes" by Shannon Othus-Gault is a product of work by Chemeketa Press.

Figure 13.2. "1969 Afghanistan (Sistan) wind ripples" by Kempf EK is licensed under CC BY SA 3.0 (https://commons.wikimedia.org/wiki/File:1969_Afghanistan_(Sistan)_wind_ripples.tiff).

Figure 13.3. "Satellite image of Parabolic Dunes" by Google Earth is under copyright and used with permission (http://www.google.com).

Figure 13.4. "Endurance Crater's Dazzling Dunes" by NASA is in the public domain (http://mars.nasa.gov/mer/gallery/press/opportunity/20040806a.html).

Figure 13.5. "Star-dune" by Elmi is licensed under CC BY SA 2.5 (https://commons.wikimedia.org/wiki/File:Star-dune.jpg).

Figure 13.6. "Desert areas of the US" by Cierra Maher is is a product of work by Chemeketa Press.

Figure 13.7a. "Satellite image of Jackson Hole" by Google Earth is under copyright and used with permission (http://www.google.com).

Figure 13.7b. "Topographic map of Jackson Hole" by USGS is in the public domain.

Figure 14.1. "Major North American Deserts" by SouthEastern Arizona Governments Organization is in the public domain (https://in.nau.edu/wp-content/uploads/sites/212/FINAL_SEAGO-2021-2025-CEDS-4.30.21.pdf).

Figure 14.2. "USGSlabels, AI upscaling: Hike395" is in the public domain and is available via Wikimedia Commons.

Figure 15.1. "Debris Flow" by Autumn Christensen is a product of work by Chemeketa Press.

Figure 15.2. "Rock Fall" by Shannon Othus-Gault is a product of work by Chemeketa Press.

Figure 15.3. "Slump" by Shannon Othus-Gault is a product of work by Chemeketa Press.

Figure 15.4. "Creep" by Autumn Christensen is a product of work by Chemeketa Press.

Figure 15.5. "Landslide overview map US" by USGS is in the public domain.

Figure 15.6. "Oso Landslide" by USGS is in the public domain (https://wa.water.usgs.gov/projects/sr530/images/figure2_remotesensing.jpg).

Figure 17.1. "Boulder measurements chart" by Cierra Maher is a product of work by Chemeketa Press.

Figure 17.2. "Flood levels photo" by Scott Burns is under copyright and used with permission of Dr. Scott Burns.

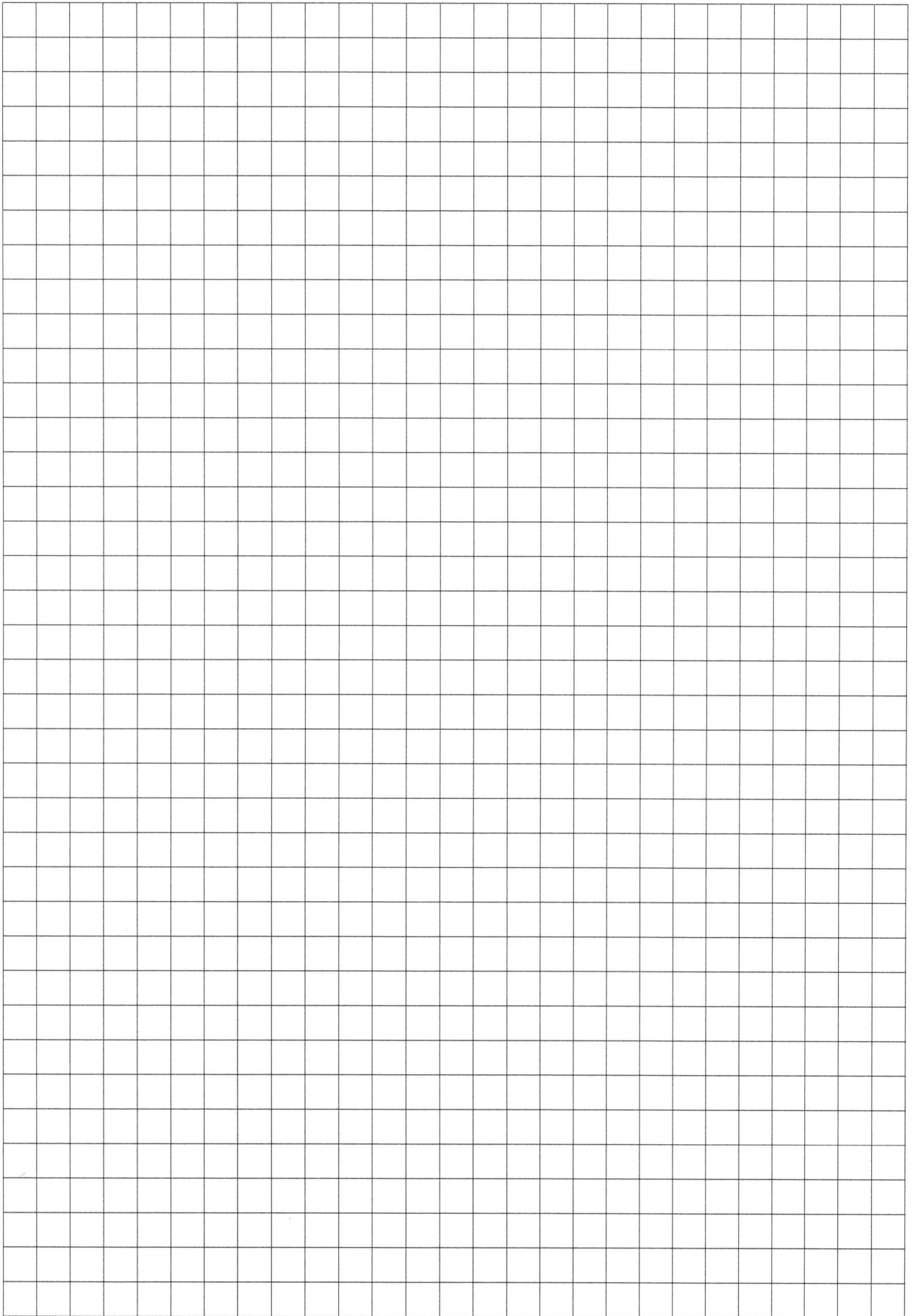

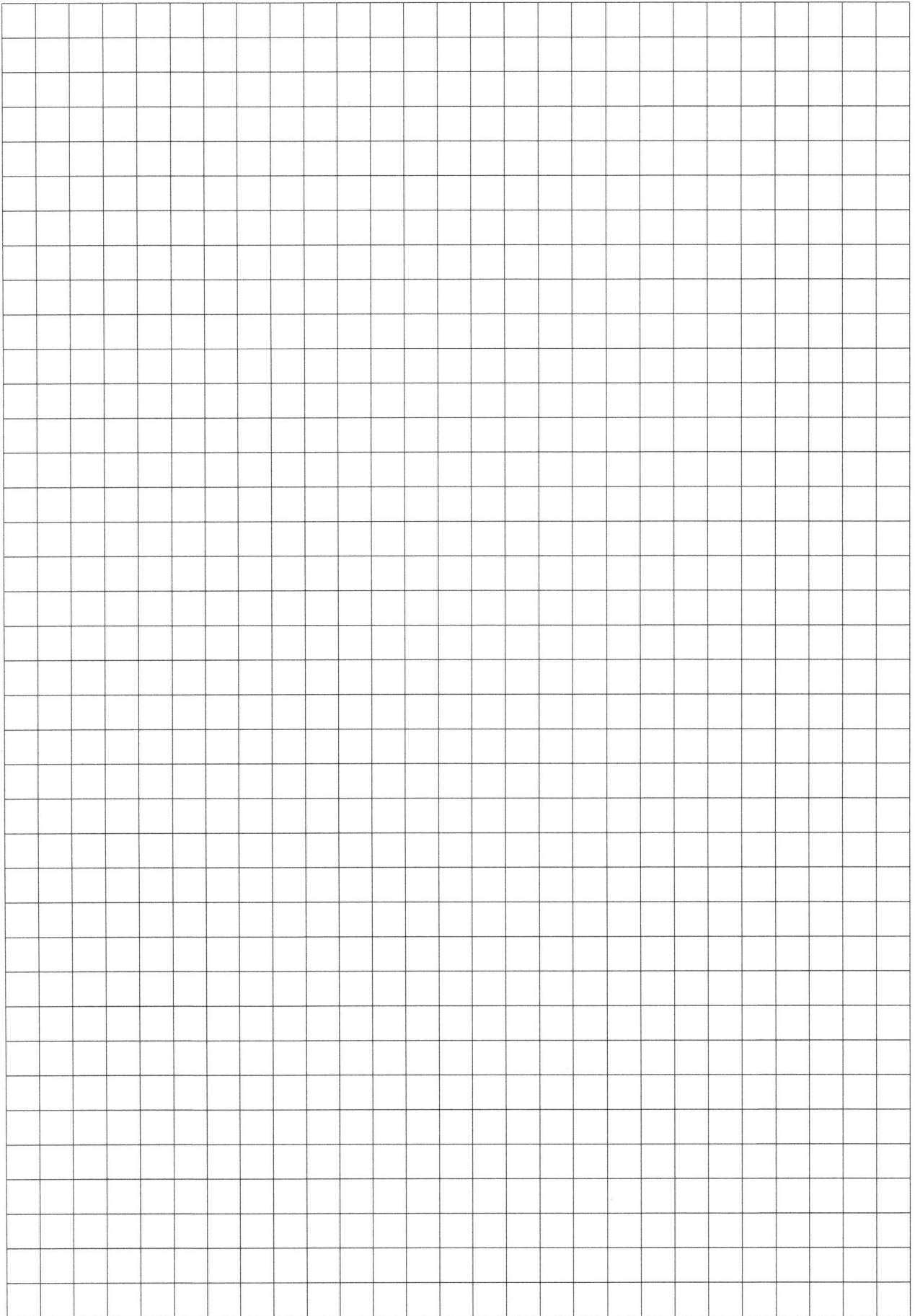

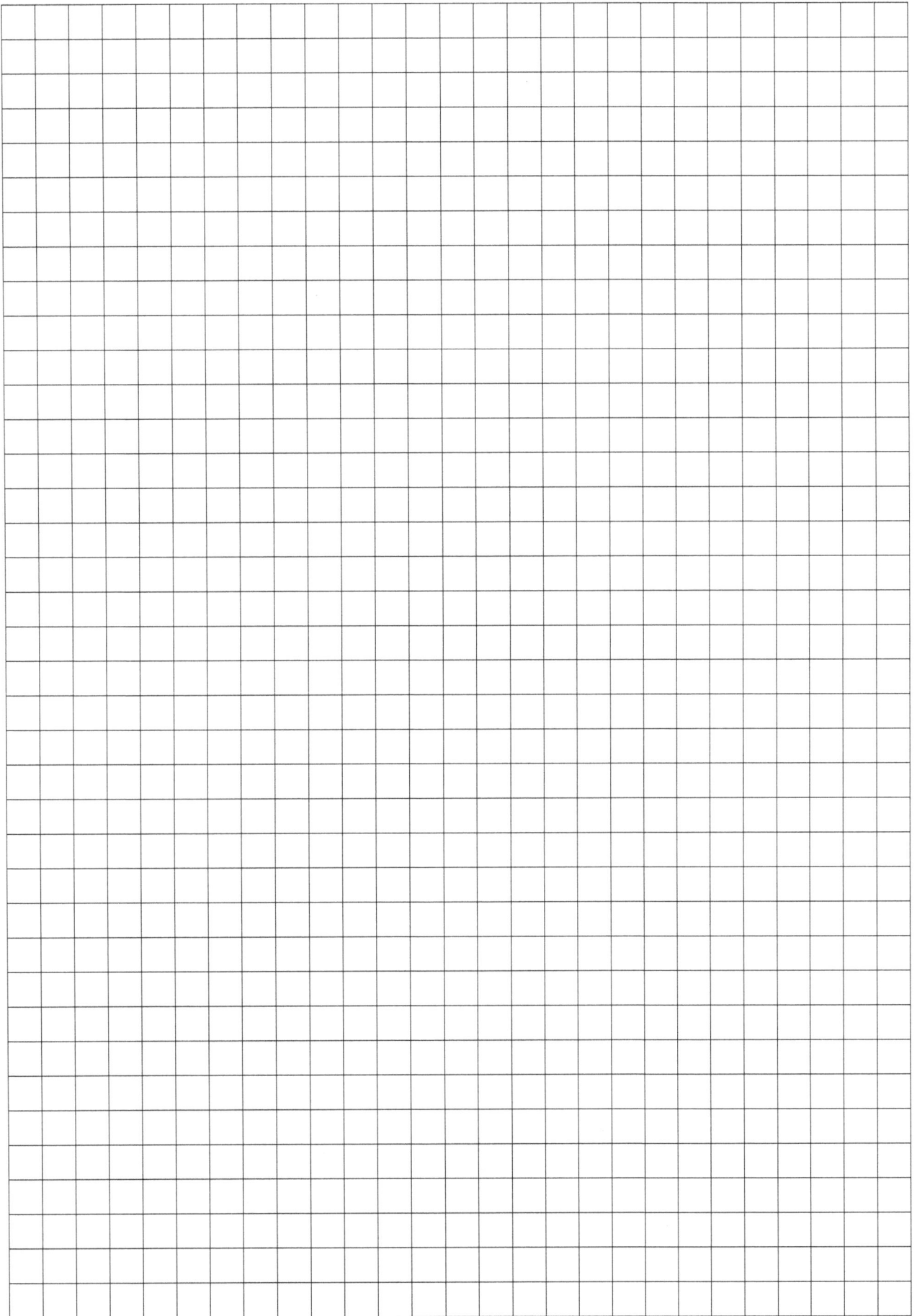

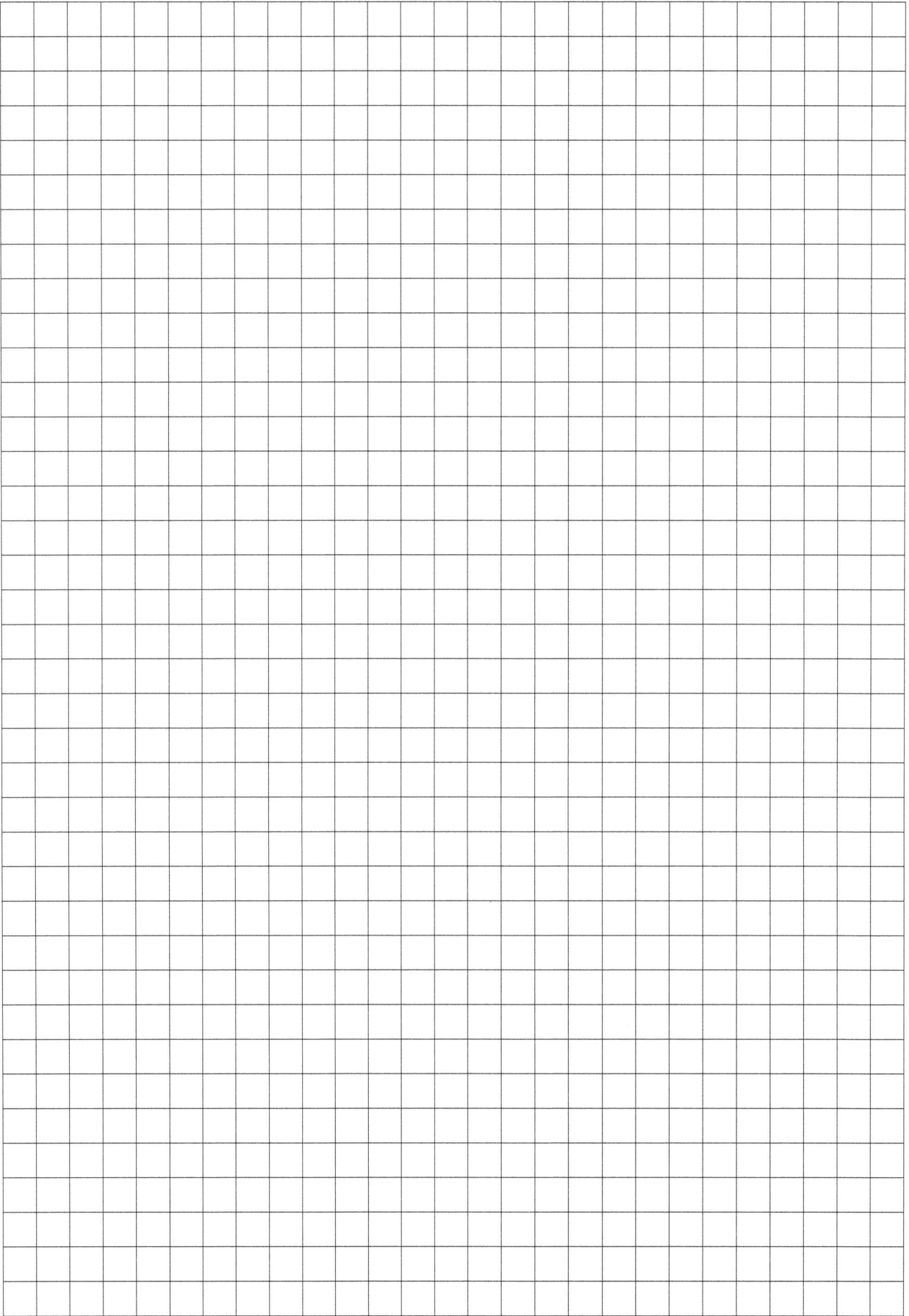

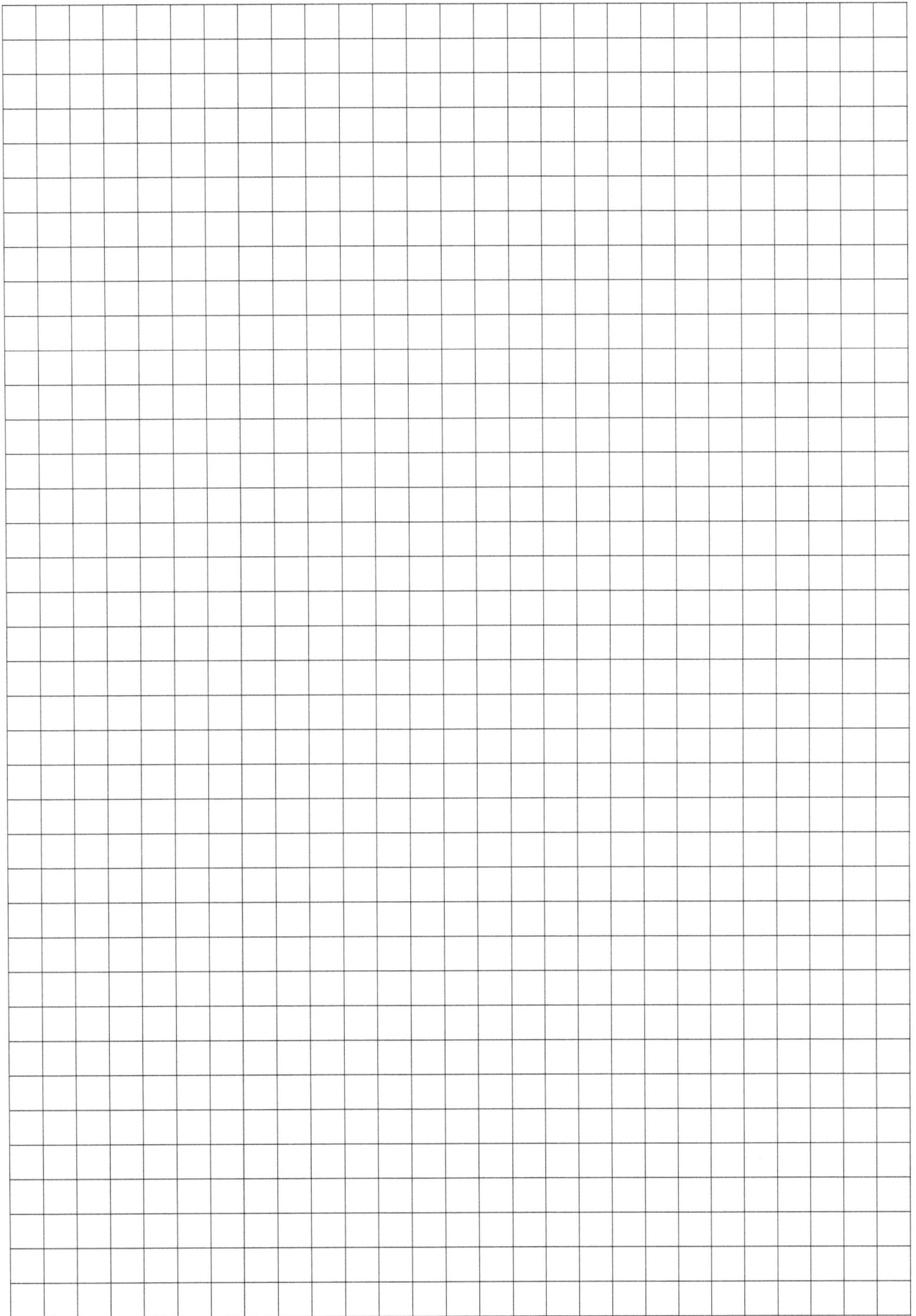

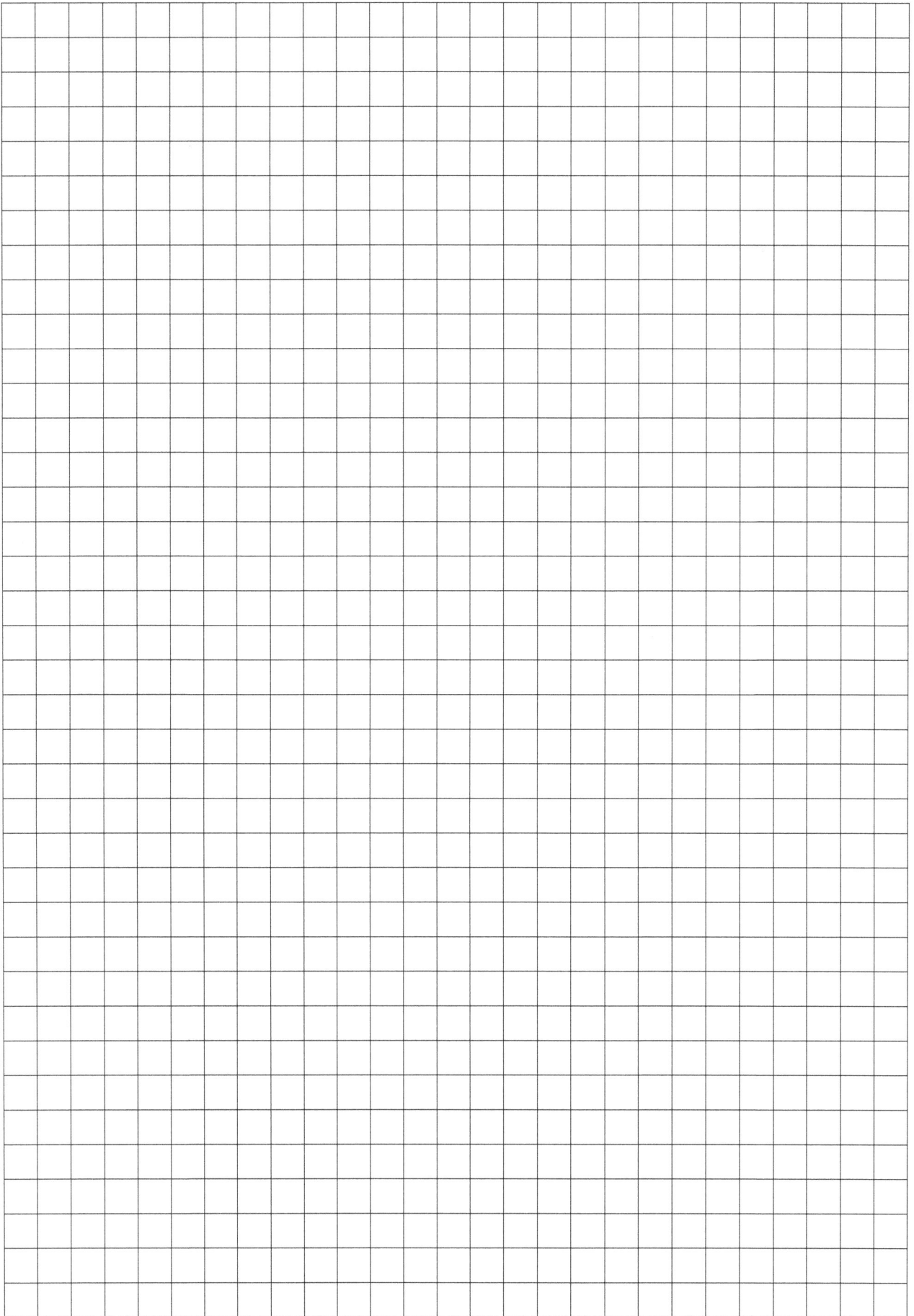

www.ingramcontent.com/pod-product-compliance
Lightning Source LLC
Chambersburg PA
CBHW081302270226
40001CB00002B/2